More Praise for *Lincoln's Last Trial*

"[The case] cemented Lincoln's image as a courtroom star–and Abrams and Fisher have made the most of their material, polishing a musty transcript into an entertaining slice of life."

—*USA TODAY*

"Abraham Lincoln was involved in thousands of cases in his distinguished legal career, few more intriguing than the 1859 murder trial of 'Peachy' Quinn Harrison… Abrams and Fisher adeptly place the Harrison trial within the context of Lincoln's legal career and his well-known skills before a jury."

—***Kirkus Reviews***

"Legal affairs journalist Abrams and coauthor Fisher illuminate a key marker on Abraham Lincoln's path to the White House… The transcripts reveal Lincoln at his best, fighting for a cause he believed in with brilliance and passion–qualities that would serve him so well as president."

—***Booklist***

"Abrams and Fisher quote generously from Hitt's transcript to bring into sharp focus the witness-by-witness testimony and courtroom proceedings."

—***Library Journal***

"Dan Abrams tells the story of Lincoln's last trial as an immersive true crime and courtroom drama experience, with impeccable research and highly entertaining digressions on such topics as 19th century jury selection."

—***CrimeReads***

"Many aspects of the Harrison trial made it spellbinding and important…. [*Lincoln's Last Trial*] adds a new layer of understanding about how Lincoln's mind worked as a consummate trial lawyer, and how that mind provided the platform for his political prowess."

—***Washington Independent Review of Books***

"A vivid portrait of Lincoln, other attorneys, what happened in the tragic case and the era.... *Lincoln's Last Trial* may represent his last major case as an attorney but hopefully it is just the beginning of a series of history books from Abrams and Fisher."

—*Valdosta Daily Times*

"You didn't know that Abraham Lincoln was the defense lawyer in a notorious murder case on the eve of his presidency? Neither did I. But Dan Abrams and David Fisher tell the remarkable tale in *Lincoln's Last Trial*, and the story is both compelling on its own terms and a lesson about some eternal truths about criminal justice."

—Jeffrey Toobin, author of *American Heiress*

"We all know the story of Abraham Lincoln the wartime president, the defender of the Union, and the emancipator of the slaves. But Abraham Lincoln, the defense lawyer? Dan Abrams and David Fisher recount the engaging story of Lincoln's last trial, occurring on the cusp of the Civil War. An entertaining book filled with twists and turns and tailor-made for Civil War buffs."

—Jay Winik, author of *April 1865* and *1944*

"Lincoln's wartime leadership overshadows his life as a lawyer. But you can't understand one without the other. In this rich and previously unexplored corner of history, the authors take you inside the courtroom to watch Abraham Lincoln—at the height of his powers as a lawyer and on the edge of eternal fame—as he tries a thrilling murder trial to a jury."

—Chris DeRose, *New York Times* bestselling author of *The Presidents' War, Congressman Lincoln* and *Founding Rivals*

LINCOLN'S LAST TRIAL

Also by Dan Abrams & David Fisher

Theodore Roosevelt for the Defense
John Adams Under Fire

LINCOLN'S LAST TRIAL

THE MURDER CASE THAT PROPELLED HIM TO THE PRESIDENCY

DAN ABRAMS

AND DAVID FISHER

HANOVER
SQUARE
PRESS

ISBN-13: 978-1-335-01562-4

Lincoln's Last Trial: The Murder Case That Propelled Him to the Presidency

First published in 2018. This edition published in 2020.

This edition published by arrangement with Harlequin Books S.A.

Library of Congress Cataloging-in-Publication Data has been applied for.

Hanover Square Press
22 Adelaide St. West, 40th Floor
Toronto, Ontario M5H 4E3, Canada
HanoverSqPress.com
BookClubbish.com

Printed in U.S.A.

I dedicate this book to my mentor and father,
Floyd Abrams, whose genuine love for, and mastery of,
history and the law established a lifelong example that I
continue to emulate but will never quite live up to.

LINCOLN'S LAST TRIAL

INTRODUCTION

Like most, my professional endeavors are largely separate and distinct from personal hobbies and passions. Never the twain shall meet, unless, of course, you are truly charmed.

The bulk of my professional career has been spent covering, analyzing and evaluating the most well-known and often notorious legal cases. Some momentous, others heartbreaking and many just driven by a well-known individual accused of some foolish act. But the best ones, the trials or arguments where I looked forward to sitting on rigid wooden benches for long stretches within the tight confines of a courtroom gallery, were the close ones. The ones where the verdict or ruling was uncertain and each day's testimony or lawyering could alter the outcome and potentially change minds, including my own.

Fortunately for me, my encounters with the criminal justice system, at least thus far, have been strictly professional.

On the personal side, about a decade ago I finally realized that rather than curling up with magazines before bed, I find it far more rewarding to have a book or e-reader in hand every night. History was always of great interest to me as a student, but recently I've become a borderline buff. World War I and II, Theodore and Franklin Roosevelt, Thurgood Marshall, Henry Ward Beecher, the Civil War, its aftermath, even James Garfield and, yes, Lincoln have been the subjects of some of my favorite nonfiction tomes. With the opportunity to tell the story of Lincoln's final murder trial, pairing my fascination with history and my work in the law, I am now officially charmed.

Even by today's standards, *The State of Illinois v. "Peachy" Quinn Harrison* would be considered newsworthy. The prerequisite elements for a "high profile" trial were all there: a well-liked and promising young man stabbed to death by a neighbor with whom he had grown up; a struggle and claim of self-defense; eyewitness testimony, including from the victim's own brother; an alleged deathbed admission; a critical celebrity witness; and a community fiercely divided. That division stemmed, in part, from the fact that this was one of those close ones where the lawyering could absolutely impact the outcome.

So if Abraham Lincoln had not been retained to represent the accused, the case still would have been closely watched and scrutinized by the community and local media alike. But in 1859, just nine months before the Republican convention that would catapult him to the presidency and eventually schoolbooks of every American child, having Lincoln for the defense made it that much more notable. While hardly a household name yet, political insiders certainly were coming to know of Abe Lincoln, particularly on the heels of the Lincoln-Douglas debates. This trial would serve as an audition for some, a second look for others, but in either case Lincoln had far more to lose than gain. The starting point for this book was a pristinely stored transcript (which very few cases had at that time) of the trial unearthed in the late twentieth century. While the discovery of this meaningful piece of Lincoln history was covered as a news story in 1989 by some of the usual media suspects, I was struck by how limited that coverage and analysis was, and has been. After all, we are talking about the final murder trial Abraham Lincoln ever tried and a truly compelling one at that. Having his exact words from a captivating trial should not be, for the most part, lost to history.

Every quotation cited from the trial comes directly from the handwritten pages meticulously recorded by this book's protagonist, Robert Roberts Hitt. There are times when the reader may want to pause, the way the jury probably did, to try to envision and comprehend the precise locations described by witnesses at the frantic scene of the incident. Ascertaining exactly who was where when must have been as challenging for jurors as it can be, at times, for the reader 150-plus years later.

To truly understand the context of the trial itself, however, we had to conduct an enormous amount of legal and historical research beyond this courtroom.

It included an exhaustive examination of how this case comports with others Lincoln argued through his career and in so doing presents a peek at the evolution of American law until 1859. In particular, the law of self-defense, which, as it turns out, was not significantly different from the standard that exists in most states today. Interspersed with the story of this trial, this book presents a snapshot of the state of the law in America in what I hope is an entertaining and edifying way.

This is, in the end, a story about a great lawyer trying a difficult case, but not in Chicago. Not in New York. From beginning to end this is about a Springfield, Illinois, attorney arguing, even performing, in his hometown. And in Springfield, Abe Lincoln was already the, well, Lincoln of his day. Every detail down to the restaurants, streets and hotels in Springfield is accurate. Descriptions of what Lincoln did and even purchased on particular days during the trial was gathered from historical data. Newspaper reporting from the time assisted with detailed descriptions of the trial beyond exactly what was said in court (which is in the transcript) on everything from the weather to the reactions from the gallery, to the challenges Springfield was facing with telegraph technology that summer. And a host of Lincoln books (cited in the bibliography) provided additional information and perspective on Lincoln the lawyer. James M. Cornelius, the curator of the Lincoln collection at Springfield's Abraham Lincoln Presidential Library and Museum, also reviewed the manuscript for accuracy.

With that said, there were times we had to deduce what was said in meetings or private conversations or suggest appropriate thoughts and/or mannerisms. We kept any re-created quotes from Lincoln himself to an absolute minimum in part because there is so much information about him readily available, and even then we have been careful to base those quotes on his previously stated beliefs as well as known conversations. We also made certain minor assumptions on the timing of the trial where the historical record was ambiguous. But every aspect of, and detail in, this story is rooted solidly in fact. The participants in the trial and dialogue from it are real. The information, history and law entirely accurate.

The trial, and the spectacle it became, is seen through the eyes of Hitt, the scribe whose handwritten words take us into the courtroom and serve as the yoke of the story. Transcribing every word of an often fast-moving trial with witnesses who did not always speak up, using just a steel-tipped pen and paper was, at times, a herculean task. Blanks and notations in the transcript reveal certain points when Hitt struggled, but on the whole, there is just no question that Hitt's work was exquisite. As we discuss in the final chapter, he (and many of the other participants from this trial) went on to an illustrious career in politics and public service. His work and contacts from this trial certainly helped pave the way. But it was his notes about the man who hired him for the job, regularly dispatched to newspapers around the country even during the trial, that helped launch a presidential candidate. "Peachy" Quinn Harrison's attorney arguing for his freedom foreshadowed many other pleas for freedom that would come later.

But we aren't there yet. To get there, we must first follow the course of a murder trial with Abraham Lincoln for the defense.

CHAPTER ONE

Mister Robert Roberts Hitt, the well-known steno man, arrived in Springfield late on the sweltering afternoon of August 28, 1859. As he stepped down onto the platform of the new station, he paused briefly and nattily patted the beads of sweat from his forehead, then vainly attempted to tug the wrinkles out of his jacket. The Alton Express had covered the two hundred miles from Chicago in a quite acceptable nine hours. Hitt had tried with limited success to practice his shorthand on the ever-shaking rails. It had not surprised young Hitt that the carriage was far more crowded than he had previously experienced: the Peachy Quinn Harrison murder trial had attracted considerably more attention than might otherwise have been expected once it became known that Abe Lincoln was going to defend the accused killer.

A fair number of people were coming to Springfield to take the measure of Abraham Lincoln. His seven debates with incumbent senator Stephen Douglas the previous fall had gained him a national reputation, and the bigwigs were predicting he was going to make a run for the presidency in 1860. There was great interest in the man, and all the regional newspapers were covering the trial, including the *Chicago Press and Tribune*. The country had been introduced to him, thus far liked what it saw, and his conduct during the murder trial might help determine if this early flirtation would turn into a serious courtship.

Hitt already knew the cut of the man. Although he dismissed it humbly when offered credit, it was his fine work that had helped bring Lincoln into prominence. As a student at the Rock River Seminary and DePauw University, Bob Hitt had taken a keen interest in phonography, the skill of rapidly converting spoken words to print, and became quite an expert at this form of shorthand. It was a well-paying trade, and he opened his own office in Chicago in 1856, working regularly for the state legislature, the courts and on occasion newspapers, which were rapidly adopting this new form of journalism. He had first met Abe Lincoln in '57, when he had been hired by the *Chicago Daily Press* to cover the *Effie Afton* trial, a landmark case that was to determine the balance between traditional river navigation rights and the construction of railroad bridges over a waterway. During that trial Lincoln had taken a liking to the energetic young reporter.

The great stir created when Lincoln and Douglas announced they would debate the complex moral, legal and economic issues of slavery and state's rights throughout Illinois caused *Press and Tribune* co-owner and managing editor Joseph Medill to hire Hitt to bring a word-for-word account of the debates to his readers. Hitt's transcriptions caused quite a stir. As newsman Horace White recalled, "Verbatim reporting was a new feature in the journalism in Chicago, and Mr. Hitt was the pioneer thereof. The publication of Senator Douglas' opening speech in that campaign, delivered on the evening of July 9, by the *Tribune* the next morning, was a feat hitherto unexampled in the West."

Hitt's transcriptions were sent by telegraph to newspapers throughout the entire country, including Horace Greeley's important *New York Tribune*, and in several weeks transformed a little-known Illinois lawyer into a widely admired political figure.

Lincoln was so taken with these transcripts that he requested two copies from the *Press and Tribune*'s co-owner Charles H. Ray, offering to pay "for the papers and for your trouble. I wish the two sets, in order to lay one away in the raw, and to put the other in a Scrap-book." That "Scrap-book," as he referred to it, was eventually published and sold more than thirty thousand copies, which added to his luster.

Lincoln spoke plainly and forcefully, the power of his words defining his character for many thousands of Americans; but he also was shrewd enough to appreciate Hitt's value to his ambitions. So much so that at the second debate, in Freeport, Lincoln was dismayed when Hitt was not present onstage and refused to begin until the steno man could be lifted out of the throng and seated on the platform, his papers on his lap.

The two men, the tall, angular Lincoln and the small and slender Hitt, had become good friends, and were brought together again at this murder trial. Lincoln had already agreed to defend Harrison when Hitt was hired by the *Illinois State Journal* to provide a daily transcript for its readers. But as the trial began both the families of the accused, Quinn Harrison, and the victim, Greek Crafton, purchased copies of that transcript, which would prove a valuable tool should an appeal be necessary, paying $25 each. Lincoln subscribed for an additional copy from his friend Hitt, for which he paid $27.50. Hitt also was permitted to provide copies of his transcript to newspapers. If that recording might also prove beneficial in spreading Lincoln's reputation, well, certainly none of his admirers would find that objectionable.

Hitt was quite pleased to accept this opportunity. In addition to his substantial fees, the attention given to the trial would also boost his own growing reputation as a pioneer in the field. Trial transcriptions were still extremely rare. There existed no devices to assist the reporter, meaning every word had to be captured by hand, an extraordinarily time-consuming and difficult process. Especially in a courtroom where complicated legal terms and phrases were proudly bandied about. And, in most instances, there didn't seem to be much need for every word to be written down. Judges didn't rely on transcripts to make their decisions. Summaries of the proceedings had long been sufficient. And when necessary lawyers were expected to be truthful in their memories.

Like most early practitioners of this profession, Hitt had developed his own system. Although relying mostly on the phonemic orthography method introduced in 1837 by the Englishman Sir Isaac Pitman, in which symbols represented sounds that could later be transformed into words, he had added those flourishes necessary for American English. He usually made his own transcriptions from those notes, but when pressed for time his assistant, a French Canadian named Laramie, would assist him with them.

In addition to a change of clothing, Hitt carried in his carpetbag the necessary tools of his trade: several of the new Esterbrook pens with long-lasting steel nibs, a supply of ink and a sufficient number of loose sheets of machine-produced paper. While the potential benefits to Lincoln's political future that could be gained by providing an exact transcript to the Chicago newspapers—and beyond—were obvious, there also was an element of risk: Lincoln's reputation was relatively untarnished; he had won the popular vote in the Senate election of '58, but at that time senators were appointed by state legislatures, and the Democratic majority in the Illinois legislature had awarded the seat to the Democrat Douglas. Should he lose the trial, should he even make a major misstep, should Peachy Harrison be convicted of murder, the spotlight now focused so brightly on him might be dimmed. Creating an aura of invincibility is the goal of every person who stands for election, and a loss in the courtroom might easily damage that perception.

As Hitt walked purposefully from the station to Lincoln's office on the west side of the public square, he noted with satisfaction how rapidly Springfield was growing. Only two decades earlier, when Lincoln had arrived there to pursue his legal career, it had been barely larger than a village. It had been settled in 1819, with thick woods to the north and a vast prairie on its south. Slightly more than a thousand residents lived mostly in small-frame houses and cabins, and farm animals roamed freely on the black mud streets. The pigs, especially, lent an unmistakable aroma that seemed to have been built into the foundations. There were no pavements or sidewalks, and large chunks of wood had been laid down to ease crossing the streets, which were, according to Lincoln's biographers Nicolay and Hay, "of an unfathomable depth in time of thaw." But even then, as Lincoln had written in a let-

ABRAHAM LINCOLN PRESIDENTIAL LIBRARY AND MUSEUM

The first railroad line in Springfield opened in 1842, though it was not permanent until a decade later. By 1856, when this photo of the railway station was taken, the Iron Horse had brought the world to Springfield, knitting it to the wonders of Chicago and beyond.

ter to a friend, it was "a town with some pretentions to elegance." It was a place with a spirit. Even in those early days Springfield was a town that rewarded enterprise, a perfect spot to settle for a bright young man with passion. Lincoln eventually fit in well there. As a member of the state legislature in 1837, by dint of his "practical common sense (and) his thorough knowledge of human nature," wrote fellow state official Robert L. Wilson, he had been instrumental in convincing his reluctant colleagues to transfer the state capital there from Vandalia.

Progress had come to Springfield slowly, but

steadily. By the early 1850s the nearly six thousand residents of this "prairie Philadelphia" were connected to the world by telegraph, the railroad and several newspapers. Impressive carriages had become common sights on the newly paved streets. The bankers had come to town, bringing with them all the necessary and desirable services and trades. Shops carried fashionable items brought from New York, Boston, St. Louis and Philadelphia. It was a prosperous city now, and politicians from the various parties, among them Douglas Democrats, Fillmore Americans and the newly formed Republicans, converged there to plot party strategy, do state business and perhaps cut themselves a slice of the public pie.

And people who needed to know were aware that there was at least one active station on the underground railroad carrying escaped slaves to freedom. Although no one dared talk about it in public, Lincoln's neighbor, a free black man named Jameson Jenkins, a teamster who lived only five doors away from the Lincoln family, was believed to be the primary conductor.

It was a fine place to be a businessman, a working man or a lawyer; especially a lawyer. As the seat of the Sangamon County circuit and the home of the state supreme court, at times it seemed to be overflowing with lawyers. Lincoln had started his career there by borrowing a horse and riding what was known as circuits, sleeping two lawyers to a bed in a tavern and killing a chicken for his dinner, but time and effort had made him among the most respected members of the local bar. Springfield was a bustling city, but not yet large enough to support all the lawyers practicing there. In 1839 the Eighth Judicial Circuit was formed, consisting of numerous contiguous counties too small to maintain a regular court presence. It grew to include fourteen counties covering ten thousand square miles. Like a traveling show or even a circus, twice a year a caravan including a judge and a pack of lawyers would bring the law to each county seat, setting up court for a period ranging from a few days to two weeks. As soon as they settled in, they would be approached by people in need of their services and to strike deals. Sometimes they would conduct the trial themselves, but equally often they would be hired to assist a local practitioner.

Like most attorneys of that period, Lincoln had no specialty; he dealt with both civil and criminal cases. When called upon he would sit as a judge, act for the community as a prosecutor or represent individuals with a gripe, a claim, a need or a criminal charge. He might just as easily be found settling a minor land claim as arguing an historically important case to determine navigation rights on America's rivers; on the same day he might write a will in the morning and defend an accused rapist in the afternoon.

While certainly it lacked the sparkle of Chicago, Hitt found Springfield to be a pleasant city where a visitor could easily find a clean place to stay, a satisfying meal and whatever entertainments were of interest.

The law office of Lincoln and his younger partner, William Herndon, was in a back room on the second floor of a brick building on the northwest side of the square, on Fifth Street. It faced the courthouse square. A small signboard on the street reading Lincoln & Herndon directed potential clients to the narrow stairway. Hitt knew Lincoln to be a man who carried most of what he needed in his mind, although when necessary he actually carried his legal papers in his tall hat, and spent as little time as possible on the mundane chores that chewed up the day. So Hitt was not at all surprised when he walked into the clutter.

It would be charitable to describe Lincoln's office as messy. It was well beyond that, well beyond anything that might be fixed by a regular cleaning. Or two. A pair of windows overlooking the backyard were caked so thickly with dirt and grime that very little light edged into the room, which actually was a sort of blessing, as the dim light obscured much of the disorder. One of the many young men who did their legal training there claimed to have found plants from a forgotten bag of bean sprouts growing in a pile of dirt in a corner. A large ink blotch stained one wall, supposedly the result of an angry law student firing an inkstand at another young man's head. One visiting attorney, Henry Whitney, wrote flatly, "no lawyer's office could have been more unkempt, untidy and uninviting than that of Lincoln and Herndon."

A long table ran lengthwise down the middle of the room, crossed at the far end by a smaller table to form a T. Both tables were covered with a faded green baize and numerous piles of papers and law books. Other piles were scattered on the floor around the office. On a stack of papers tied together with a string Lincoln had written boldly, "When you can't find it anywhere else, look in this."

The afternoon had begun cooling into the early evening, but Hitt was not at all surprised to find Lincoln still at work; it was well-known that he found Mrs. Lincoln to be difficult, and he often found peace in his office. Besides, with the Harrison trial about to begin there was considerable preparation to be done. As Hitt entered the office, he heard Lincoln reading out loud from his notes, as was his practice, then saw him stretched out full-length on two hard chairs, his eyes fixed on the pages.

"Mr. Hitt," Lincoln greeted him, straightening up. "Thank you for coming."

"Mr. Lincoln," Hitt responded. "And thank you for arranging it." Hitt cleared another pile of books from a chair and sat down. The two men exchanged pleasantries. While Lincoln seemed to be as naturally engaged as in their previous meetings, Hitt was surprised to feel a distance between them, as if he were seeing an old friend through new eyes. Odd, he realized, all the hoopla around Lincoln appeared to have affected him to a far greater degree than it had the man himself. He realized with a start that he was responding more to the created image than the flesh and blood person. But there was nothing to be done about it; the public had its hooks into him.

Lincoln did not pretend to be unaware of his changed circumstances; he had been pursuing public office for two decades, and his efforts were finally coming into bloom. It was what it was, and there was no reason now to ignore that reality. He was receiving numerous requests to speak at political gatherings around the country, even as far away as New York City, but most of them he had turned down.

What daylight that somehow had managed to seep through the windows gradually disappeared. Lincoln lit two candles and began telling stories. This was the man his companions knew well, a man always ready with a story, often told at his own expense. Hitt had heard it said that no one knew the real Abraham Lincoln—he kept much of what motivated him safely boxed inside—but everybody knew his stories. They were almost legendary, and those he didn't repeat personally others did for him. Everybody who knew Lincoln had heard him explain that while he had served in the Black Hawk War, "If (General Cass) saw any live fighting Indians it was more than I did; but I had a good many bloody struggles with the mosquitoes..."

His ability to capture a room with some soft-spoken words had served him well, whether sitting around the evening fire in the back of his friend Joshua Speed's general store or transfixing a crowded courtroom. Hitt had asked him about the state of his practice, which after several turns led him to a story about a recent murder case that had raised some questions. As he told it, after suffering through half a century of a bad marriage, seventy-year-old Melissa Goings had finally reached her limit. When her seventy-seven-year-old husband, by all accounts an irascible old codger, had tried to choke her, she took hold of a solid piece of stove wood and hit him twice, fracturing his skull and killing him.

The old man wasn't much missed, but the law had little sympathy for his widow. Lincoln's face glowed in the flickering candlelight as he continued his tale. The widow Goings was indicted and charged with her husband's murder, he continued, and some sympathetic townsfolk hired him to defend her. The facts were not in her favor, and the judge was known to be tough. On the morning her trial was to begin, Lincoln received permission from the court to consult privately with his client on the ground floor of the courthouse. "I reckon I did just that," he told Hitt. But later, when the bailiff came to get her, Melissa Goings could not be found. Asked by the judge where she had gone, Lincoln recalled, he had shaken his head and replied, "Don't know, your honor. I left her on the lower floor. She should be in the care of the sheriff."

Well, Lincoln continued, Melissa Goings apparently was gone, never to be seen again. "The bailiff, Mr. Robert Cassell, was plenty upset with me. He said strongly, 'Confound you, Abe, you have run her off.'

"'Oh no, Bob,' I said to the man, 'I did not run her

off. She just wanted to know where she could get a good drink of water—'" here he paused for the maximum effect "'—and I told her there was mighty good water in Tennessee!'" There was a sprightly joy in his voice as he delivered his final line, as if he were loudly sharing a secret known by everyone in the world. It would be rumored later that he had gone so far as to open the window for her, and helped her climb out of it, but that was never proved; in fact, there were never any claims of wrongdoing of any kind. The case was dropped. While clearly Lincoln had stretched the limits of the law, it was just as clear that he believed justice had been served, if not quite done.

Hitt was worn from his long trip, and hungry, and he hadn't yet put his bag down in the rooming house, but there was time for all that. He was not about to give up these relaxing minutes with Lincoln. And, as he expected, eventually Lincoln came around to telling him about the case at hand.

Hitt knew the general facts—the newspapers had been writing about them—but Lincoln added the color. Seven weeks earlier, on Saturday the sixteenth of July, Greek Crafton had walked into the Short and Hart's drugstore in Pleasant Plains, Illinois, about ten miles outside Springfield, carrying with him a ton of anger. His older brother John Crafton was stretched out on a counter; his reason for waiting there was still in dispute. Simeon Quinn Harrison, "Peachy" as he was called, was sitting at the other counter next to Mr. Short, reading the newspaper. Harrison was a frail young man, weighing no more than 125 pounds, and Greek had a substantial advantage of size over him. Greek took off his coat, and with his brother grabbed hold of Harrison and pulled him away from the counter. According to Lincoln, the Crafton brothers tried to drag Harrison into the back of the store to administer a sorrowful thrashing. Proprietor Benjamin Short tried to step between the boys but was pushed away by brother John.

The smaller Harrison struggled to break free. Greek Crafton struck him a hard blow. The brawlers fell over a pile of boxes. With that, Harrison pulled a four-inch-long white-handled hunting knife and began slashing at his tormentors. Stabbing wildly, he made a deep slice into Greek's stomach. As Greek stumbled away, Harrison stabbed at John, catching him on his wrist and opening a nasty cut. To keep Harrison away, John threw a balance scale, some glasses and even a chair at him before they finally could be separated.

But Greek had been grievously wounded, cut from the lower rib on his left side to his groin on the right. Crafton's bowels protruded and later were pushed back into place by Dr. J. L. Million. He was taken to bed and lingered for three days, but there never was any real hope of survival. Supposedly, on his deathbed, as Greek prepared to meet his Maker, he wanted to set himself right. Peachy's own grandfather, the renowned Reverend Peter Cartwright, had gone to pray with Greek, and was stunned when Greek said to him, "I brought it upon myself and I forgive Quinn, and I want it said to all my friends that I have no enmity in my heart against any man. If I die I want it declared to all that I died in peace with God and all mankind." But a statement of forgiveness wouldn't change the law.

Quinn was arrested and his family hired the men who would defend him.

The event inflamed local passions. Everybody in the area knew the boys, Quinn Harrison and Greek Crafton, and held an opinion about them. "There is much excitement manifested among the friends of the respective parties," the *Daily Illinois State Journal* reported. Pleasant Plains was a small village, and to some degree each of its seven hundred residents had a stake in the outcome. Anger took hold. Loud accusations and countercharges were made as supporters of both the accused and the victim rallied behind their chosen side. The local newspapers reported the facts—but also took the position certain to appeal to their readers. "We are informed that the Grand Jury have found a bill against young Quinn Harrison for the murder of Greek Crafton," wrote the *Journal*. "By what mode of precedence they made up their minds to such an indictment is a mystery. If there ever was a case of killing in self-defense, we think the testimony at the preliminary examination of Harrison showed one."

The rival *Illinois State Register* claimed the ethical position, declaring, "Whether rightly or not, it is not for the public press now to express an opinion, in justice to either party... The *Journal*'s paragraph is one which cannot but be depreciated by every thinking man. It is unjust to the party it assumes to defend, who, if not indictable for murder, cannot be benefited by mingling his cause with the partisan commentary of the newspaper press...we protest against the conduct of a public journal which assumes to prejudge judicial investigation..."

But there was far more to the story, Lincoln continued, creating a fuller picture of this sad affair for Hitt. The village of Pleasant Plains was close enough to Springfield to almost catch its shadow; on a dry day it was an easy ride from there into the city. Lincoln had known both young men for a considerable time, he told Hitt. Both of them came from wealthy and well-positioned farming families in the community. Peyton Harrison, the accused killer's father, was a staunch Republican and a longtime friend and Lincoln supporter. The victim, Greek Crafton, also was close to him. In fact, he had aspired to the law and did much of his training as a clerk right there in the office of Lincoln & Herndon, acquitting himself quite well. To complicate matters even further, Lincoln had a long and mostly contentious relationship with Harrison's grandfather, the Reverend Cartwright, who had recounted Greek's deathbed comments. Peter Cartwright was an illustrious prairie preacher, famed for bringing the Methodist doctrine to the frontier—whatever it took. Once, supposedly, that required brawling with the legendary riverman Mike Fink.

Like Lincoln, Cartwright's power was in his words, and it was said his booming voice could "make women weep and strong men tremble."

People said about Cartwright that "When he thought he was right no earthly power could persuade (him) to abandon a principle," while adding that he always thought that he was right. There was no pretense between Lincoln and Cartwright; the two men did not take kindly to one another. Twice Lincoln had gone up against him for election. In 1832, in Lincoln's first attempt to win public office, the good Reverend Cartwright had defeated him for a seat in the Illinois state legislature. They met a second time in the congressional election of 1846, an especially nasty campaign. Running as a Whig, Lincoln objected strongly to Cartwright's insistence on bringing his religion into the public square. The Democrat Cartwright responded by tarring Lincoln as "an infidel," a man unfit to represent good Christians. Lincoln had won that election, and neither of the men had seen fit to apologize.

The two men were preparing to meet once again, but this time in a courtroom, this time on the same side, this time with a young man's life in balance. The trial might well turn on Cartwright's recitation of Greek Crafton's dying words—no juror with a heart could hear those words and not be affected—but getting them to be heard was the problem. Lincoln had to convince the judge to allow this hearsay evidence to be admitted. That was tricky territory. Hearsay is a statement made outside the courtroom by someone other than the witness. The witness, under oath, repeats what he was told or heard. When it is intended to prove the truth of the comment, it's generally not permitted in a courtroom because the opposing side cannot question the person who made that statement. But in certain specific circumstances it is, and was, allowed.

Lincoln had tried more than two thousand cases, both civil and criminal, and in many of them he had known at least one of the participants, sometimes both. But he couldn't recall a single one of them that was so entwined by personal complexities: he was charged with defending the accused killer of a young man he had admired, and in doing so he would be allied with a man with whom he had been at odds for decades.

Turns out, the incident had been sparked at least two weeks before the fatal confrontation, while the village celebrated the Fourth of July. Harrison and Crafton had known each other for many years, both had grown up in the village, for a time they had even been friends, but no one could state with certainty the precise cause of the bitterness between them. There was some talk that it concerned some difficulty over a Harrison woman. Greek's brother William had married Peachy Harrison's sister Elizabeth Catherine and there were rumors about abusive behavior, which may well have played a role. At the town picnic on the banks of the Sangamon River, Peachy Harrison had warned his younger brother Peter that he best stay away from the Crafton family, disparaging them as "not fit associates." Others claimed that Harrison had slapped Crafton, but there was not a lot of support for that. Whether Harrison did or did not lay a hand on him, Greek took umbrage at the insults making their way through the local rumor mill and challenged the smaller Harrison to a fight. It had become a matter of honor. Friends stepped between them before the situation burst out of control, but Crafton was not satisfied. Witnesses heard him threaten to whip Harrison the next time he saw him, to which Peachy warned that if Greek ever dared lay a hand on him he would defend himself—with a gun!

Lincoln paused here and shook his head sadly at the folly, knowing this waste of a good life might so easily have been avoided at so many points if only some cooler head had stepped between them…

Hitt listened silently in the flickering candlelight, recalling a story he had heard only months before, about how Lincoln himself had responded years earlier when finding himself in a somewhat similar position. In that situation, as in this one, honor had been at stake. It was impossible to know how much of the story was true and how much was shine for political purposes, but certainly there was some truth to it. During a dispute over banking policy in 1842, Lincoln and Mary Todd had written anonymous letters to the local newspaper ridiculing her former fiancé, Illinois state auditor James Shields. They called him "A fool as well as a liar." When Shields learned the authors' identities, he demanded a retraction, and when it was refused, he challenged Lincoln to a duel. They met at a place called Bloody Island, but rather than pistols Lincoln declared they would fight with large cavalry broadswords. Supposedly the much taller Lincoln intimidated his rival by slicing an overhanging branch from a tree with a single blow, causing Shields to agree to an offered truce, his honor intact. Although an alternative ending claimed mutual friends had interceded to prevent bloodshed. Whatever the actual resolution, the foolishness had been stopped without injury. But not in this case. Rather than simmering, Lincoln said, the dispute began to boil. Fearing Greek Crafton, Harrison borrowed a knife and carried it with him always. During the next few weeks, additional insults were exchanged and the threats escalated, until the fateful morning of the sixteenth.

Greek Crafton had lingered for three days before succumbing to his injuries. For a time it was also feared that his brother John, whose wrist was sliced open, might also die. He survived. Then, for almost a week following the deadly encounter, Peachy Harrison could not be found. In fact, as it became known later, he was in hiding, perhaps in fear of Greek Crafton's posse of friends or supporters, perhaps to keep himself out of the sheriff's hands until a proper defense could be arranged. For a time his good friend, the twenty-nine-year-old city attorney Shelby Moore

During his presidency Lincoln described Stephen Trigg Logan as "one of my most distinguished, and most highly valued friends." The fact that Logan, perhaps the most respected lawyer in the region, had selected young Lincoln to be his law partner provided a tremendous boost to Lincoln's career. One reason for the selection, he later wrote, was that Lincoln had the ability to bond with all people: "Lincoln seemed to put himself at once on an equality with everybody—never of course while they were outrageous, never while they were drunk or noisy or anything of the kind." Logan was one of the few people the president invited to travel with him to Gettysburg for the dedication of that battlefield.

Cullom, gave him cover in his home, then hid him under the floor of the Illinois State college building.

The boy's father, Peyton Harrison, retained an eminent former judge, Stephen Trigg Logan, and Lincoln to represent his son. There were strong bonds between all of these men. Lincoln had first represented Peyton Harrison in a minor dispute more than two decades earlier, and they had remained friends since. The wealthy Harrison had been a key contributor to Lincoln's political ambitions. Judge Logan had been a mentor to Lincoln, as well as a friend. In 1841 Logan had offered a partnership to Lincoln. It had been dissolved after three years because Logan wanted to work with his son, and by then Lincoln felt ready to establish his own firm. The partnership had been an amiable and productive relationship, and Lincoln always spoke fondly of the man and all he had learned from him. They separated quite amicably, and both men had continued to grow in stature through the years. Few lawyers in the area were more respected than Logan and Lincoln, Hitt knew, and Quinn Harrison could not have better representation. *If my fate was to be decided in a courtroom*, Hitt thought, *these are the men I would choose to defend me.*

Lincoln's current partner, Billy Herndon, himself a fine attorney, was added to the defense. And finally they accepted Harrison's suggestion that his friend Shelby Moore Cullom assist them. "My duties," Cullom later remembered, "were looking up authorities and testimony and generally doing the common work of the concern."

The coroner's inquest into the facts of the encounter, to determine if a crime had been committed, had convened on August 2. The legal question here was simply whether Greek's death should be deemed a homicide rather than some other noncriminal cause of death. Because of his prior work with Lincoln, Hitt had followed the events as much as possible in these earliest stages, although the details had been sketchy. The courtroom had been densely crowded but still the spectators exhibited fine behavior throughout the two-day hearing. As many as seventy-five witnesses were subpoenaed, and most of them eventually testified. "Listen to this, Bob," Lincoln said, withdrawing a newspaper from the pile he had been reading. "Here is the crux of the matter. This from the *Register*, 'In the most debatable testimony, Mr. Cartwright testified that Crafton, on his deathbed, absolved Harrison from blame, and blamed himself for the difficulty and its sad result… This was rebutted by Dr. Million, who stated that he had several conversations with Crafton on his dying bed, relative to the difficulty, and that he did not absolve Harrison from blame, but censured him.'" He laid down the newspaper, squeezed the bridge of his nose between his thumb and forefinger, and continued. Hitt forced himself to hide a yawn, the heat and the long trip finally taking a toll.

Lincoln explained that he and Logan had agreed immediately on their strategy for this hearing: this was clearly a case of self-defense. Peachy Quinn had been attacked by the Crafton brothers and had been forced to fight back to save himself. One witness even testified that he had heard Crafton boasting that he intended to throw down Harrison and stomp on his face.

The prosecution disagreed, of course, examining the same facts and arguing quite a different conclusion: the laws concerning self-defense had been mostly settled; a man had no legal right to stand his ground but to save himself from imminent and serious bodily harm or death—then and only then did he have the right to use deadly force. Harrison could have avoided this fight, but instead had armed himself with a deadly weapon that he was prepared to use. When given the opportunity, the prosecution argued, he had knowingly murdered Greek Crafton.

The whole of the event was argued back and forth with great skill, although Lincoln admitted to Hitt that neither he nor Logan anticipated it would end at the initial pretrial hearings. Feelings were far too raw for a decision to be reached without all of the evidence being presented and all of the witnesses heard. Failing that, the town might never heal. There would have to be a full-blown trial.

The coroner's inquest came a few weeks before the Sangamon circuit Grand Jury would officially determine if charges were to be lodged against Harrison.

That was but a formality, Lincoln knew. The Grand Jury would indict and Peachy Quinn Harrison's trial would begin within days after that. "That's the sum of it," Lincoln said finally. He then offered to escort Hitt to the Globe Tavern, the boardinghouse where he would be lodging during the proceedings. Mr. Harrison had arranged these accommodations, Lincoln told him, and he was confident the stenographer would be quite comfortable there. "I know the place well," he said. "My own family lived there for a time." It was, in fact, the first home he shared with Mary Todd; they had paid $8 a month to live there through the first year of their marriage. In fact, it was there that their son Robert had been born.

Lincoln did not bother locking his office. It would have served no purpose, Hitt noted; the two lower panes of glass in the door were missing, and he saw no effort had been made to replace them. The two men strolled slowly through the gaslit streets, Lincoln nodding to passersby as they greeted him. The heat had subsided only slightly. As they walked into the night it occurred to Mr. Robert Roberts Hitt that Abe Lincoln, a man in his ascendancy, seemed in no hurry to get home.

CHAPTER TWO

After confirming that the quiet Bob Hitt was safely installed at the Globe, Lincoln continued his walk. The steno man was right; he was in no hurry to be home. He savored these long walks, which had become less frequent as his star rose, for the moments they gave him to think. At times he became so entangled in his thoughts that he completely forgot his destination. He wandered, but used the time well.

This evening those thoughts were focused with precision on the coming trial. It was going to be a tricky business and would require all the skills he had gained in his decades of practice to bring it to a satisfactory conclusion. While he mourned for Greek Crafton, for all the possibilities that had been lost, he also believed without doubt that Quinn Harrison had acted in self-defense. It was widely reputed that Abraham Lincoln refused to defend legal positions in which he did not believe, regularly turning away tainted clients who wanted him to mold the law to fit their needs. As he once avowed, "No client ever had money enough to bribe my conscience, or to stop its utterance against wrong and oppression." While Lincoln busily prepared for this trial, in other places around this country the forces were gathering that eventually would result in his election as the sixteenth president of the United States.

His strength and his sacrifice became the stuff of legend, but the fascinating legal career that prepared him to make those momentous decisions is often ignored or forgotten. Lincoln was one of the giants who literally set the bar for the legal profession in America. By the time he walked into the Springfield courthouse the first week of September in 1859 to defend Peachy Quinn Harrison against charges of murder, he was among the most respected and experienced attorneys in the west. He had been in practice for more than two decades; he had tried thousands of civil and criminal cases covering an extraordinary range of complex issues.

In one case he argued against the right to sell a slave in Illinois, but in the controversial 1847 Matson trial, he defended a Kentucky planter (Robert Matson) living in Illinois, who was sued by his own slaves for their freedom under that state's comparatively liberal laws; Lincoln lost. He appeared before the Illinois Supreme Court more than three hundred times, and even traveled to Washington to argue a case in front of the United States Supreme Court, losing on complex technical grounds an 1849 case about the application of a state's statute of limitations to a non–state resident. Although admittedly he saw abuses of the system, he remained extraordinarily proud of his profession. "Law is nothing else but the reason of wise men," he had written, "applied for ages to the transactions and business of mankind." And he reacted strongly when people doubted the honesty of lawyers, writing in a well-known 1850 essay, "There is a vague, popular belief that lawyers are necessarily dishonest… Let no young man, choosing the law for a calling, for a moment yield to this popular belief. Resolve to be honest at all events; and if, in your own judgment, you cannot be an honest lawyer, resolve to be honest without being a lawyer. Choose some other occupation, rather than one in the choosing of which you do, in advance, consent to be a knave."

Honesty, above all, became his hallmark. According to folklore he acquired the nickname "Honest Abe" while working as a New Salem store clerk when he walked several miles to return a few pennies due a customer. But his commitment to the truth resonated throughout his entire career. It was a reputation he carried with him into every courtroom; judges, fellow lawyers and jurors came to know the value of his word, and there is little doubt that served his purposes.

In the ensuing decades, the trappings of the law have changed substantially, pushed by the changing winds to forge new paths, but its original course remains instantly recognizable. The American legal system of the 1850s generally followed the traditional format of British law and adhered to the basic rules of evidence; it guaranteed the accused a fair trial before a jury of his or her peers, and generally required unanimous agreement for a conviction. The settings might have been less formal, spittoons were a necessary feature, for example, and the language was sometimes coarse, but an attorney sliding through time from then to now, and even going the other way, would not find it difficult to understand and even participate in the proceedings.

Among the greatest differences, though, is how a man earned his way into a courtroom. The law was considered an honorable trade; a man trained for it like any other trade and didn't necessarily need schooling. There isn't any specific time or event historians can point to and say, right there—that's when Abraham Lincoln decided to become a lawyer. In fact, there was little in his childhood to lead him in that direction. In the early 1800s the legal system was loosely structured, with rudimentary courtrooms being set where space could be cleared, often in a tavern or a boardinghouse. The circuit court in rural Montgomery County, Illinois, for example, where Lincoln occasionally practiced, met regularly in a farmhouse bedroom, with the judge sitting on the bed.

Like most ambitious young men of that time, it appears Lincoln had spent at least some time inside some of those courtrooms. A trial was also considered entertainment and on "court day," as it was known, the galleries often were packed as spectators crammed inside hoping to hear great oratory.

His father, Thomas Lincoln, was involved in several lawsuits, among them a dispute involving the title to some land and an unpaid debt. When twenty-two-year-old Abe Lincoln moved from the backwoods into the settlement of New Salem, he befriended the local justice of the peace, a rotund man named Bowling Green, and continued attending legal proceedings. Some of the locals began asking him for advice and assistance. He began his career by acquiring a book of common legal forms, including deeds, receipts, bills of sale and wills, and acting as the "next friend" for his neighbors. On occasion he took a seat in the jury box. At least once he served as a character witness, although it is doubtful he did much good. Asked about the word of a man named Peter Lukins, Lincoln responded honestly, "Well, he is called Lyin' Pete Lukins."

His first real experience as a lawyer turned out poorly. When a general store he had purchased in partnership with a friend in 1833 "winked out," he was sued by his creditors. After the local justice found Lincoln and his partner liable for the debt, they appealed to the circuit court where that ruling was confirmed.

He tried his hand at several other professions before settling on the law. He was a failed candidate for the Illinois state legislature, a store clerk, worked on riverboats as a deckhand and a captain, he was a postmaster and a shopkeeper, a farm laborer and a surveyor, but the law finally caught his fancy and he committed himself to it.

When asked years later for advice on how to enter the profession, he wrote, describing his own path, "If you are resolutely determined to make a lawyer of yourself, the thing is more than half already done. It is but a small matter whether you read with anybody or not. I did not read with anyone. Get the books and study them till you understand them in their principal features…" Those books, he added, should "Begin with Blackstone's *Commentaries* and after reading it through, say twice, take up Chitty's *Pleadings*, Greenleaf's *Evidence* and Story's *Equity*, etc., in succession.

"Work, work, work is the main thing."

The only requirement for a license to practice law was a certificate granted by a court that the applicant was a man of good moral character. The Circuit Court of Sangamon County certified Abraham Lincoln as "a person of good moral character" on March 1, 1837; he raised his right hand and took an oath to support the Constitution of the United States, and Illinois; several months later his name was published in the annual roll of admitted attorneys, and he was permitted to practice.

Armed with this credential, Lincoln literally rode into Springfield on a borrowed horse, set down his saddlebags in the bedroom above Joshua Speed's store and set to lawyering. John Todd Stuart, with whom he had served during the Black Hawk War, invited him to form a partnership. Stuart and Lincoln practiced general law, meaning doing pretty much whatever legal work was needed. Lincoln learned by doing: he defended debtors and pressed debt collections; he created and dissolved partnerships, incorporated companies and declared bankruptcies; wrote and probated wills; he defended petty thieves and men charged with brawling and battery. He won some and lost others, but slowly built a practice. While these cases mostly lacked the substance and impact of those seen in New York and Boston and Philadelphia, Lincoln and men like him brought the structure and accountability provided by the law to the west.

Lincoln and Stuart's second-floor office was conveniently located directly over the county courthouse. That proved fortuitous one evening when a friend of Lincoln's, E. D. Baker, was making a passionate political speech during which he lambasted the Democratic Party and the newspapers that defended its corrupt practices. Lincoln was watching from above, lying on the floor of his office and peering through a trapdoor.

Baker's repeated insults finally brought the crowd

to its feet—ready to attack him. As several men moved forward, they stopped suddenly as two long legs seemed to drop through the ceiling—followed smoothly by all of Lincoln's six-foot-four-inch frame. Lincoln took a stand between the men and Baker, warning, "Mr. Baker has a right to speak…and no man shall take him from this platform if I can prevent it."

No one doubted either his resolution or his ability to fulfill his threat.

Stuart was a fine mentor but his heart was in politics, and he often was gone. Having lost an election for Congress in 1836, he ran again for the seat in 1838, this time defeating the Democratic candidate, the lawyer Stephen A. Douglas, by thirty-six votes of the thirty-six thousand cast. He went off to Washington, leaving the practice to Lincoln's increasingly capable management. Douglas, it turns out, eventually courted Stuart's lovely first cousin, Mary Todd, in direct competition with Lincoln.

When Lincoln's partnership with Stuart ended in 1841, he joined Stephen Trigg Logan, a former circuit court judge and perhaps the most respected lawyer in Illinois. It was Judge Logan who had signed the court register welcoming Lincoln to the profession. He had sat in judgment of several of the novice lawyer's early cases. And the two men had previously worked together in the first of Lincoln's murder trials, the 1838 shooting of Dr. Jacob M. Earley by Henry Truett.

It was in his defense of Truett that Abraham Lincoln made his first deep mark on the public consciousness.

There seemed to be little question about the facts: Earley had been sitting in the parlor of Spottswood's Rural Hotel when Truett entered and, to avenge a perceived insult, pulled his pistol and shot him. Earley lingered three days and before he died named Truett as his killer.

The defense team included Stuart, Lincoln and Logan. Ironically, the acting prosecutor was Stephen Douglas, also the first of his several legal encounters with Lincoln. Much like the Harrison case decades later, this crime polarized Springfield. As the *Journal* reported, "The gloom which this occurrence has thrown over our community can with difficulty be realized by those who have not witnessed it." There were several suggestions of a motive, but political differences clearly were at the core. Truett believed Earley had authored a Democratic Party resolution highly critical of his appointment as a Register of a Land Office in Galena, Illinois, which had led to his being replaced by Stephen A. Douglas. The trial proceeded as expected, and the evidence against Truett piled up. The prosecution introduced a motive, produced evidence that he planned the crime and convinced the judge to admit Earley's dying declaration. The outcome seemed certain—until summations began. Lincoln rose from the table and faced the jury.

Even then lawyers were showmen, and the courtroom was their stage. The favored style of the day was that of the greatest stage actors of the time, loud and broad; lawyers thundered and raged, they whispered and wept, they waved their arms and pounded the table. They performed, and a good verdict was their applause.

But Lincoln brought a different method into the courtroom. He wasn't much given to histrionics. Even though he towered above most jurors, it didn't seem like he was looking down on them. When he approached the jury box and leaned over close, he was just talking to some friends.

Eventually Lincoln would become renowned for the power of his final arguments, possessing the rare ability to move juries in their minds and their hearts by weaving facts and emotions into a plausible tale. But even in those early years, his stirring use of common language allowed him to forge a remarkable connection with his audience. People simply liked him. It was never an act; there was nothing flamboyant or showy about the man. They liked his folksy approach and his slightly disheveled appearance, he reasoned with them calmly and treated them with respect and so they believed him.

There was no time limit on the length of a summation, which fit his presentation well. He was never in a hurry when justice was at stake. In this case he laid out a different sort of scenario than that of the prosecution, showing jurors a different way of looking at the same facts, suggesting to them the possibilities of following an alternative path.

Lincoln had known the victim; in fact, he had served under Jacob Earley's command in the Black Hawk War. This incident was a sorry affair, no doubt about that, he undoubtedly told the jury, but insisted his client had acted in self-defense. The victim had been holding a chair, Lincoln argued, that became a potentially lethal weapon.

The jurors went upstairs to the Stuart and Lincoln office to deliberate. After five days of mostly damning testimony, they needed only three hours to reach a verdict.

Not guilty.

Lincoln was given substantial credit for that outcome, which greatly improved his professional prospects. His practice continued to grow steadily. When his association with Logan ended amicably after three years, Lincoln felt ready to hang his own shingle. Logan agreed, later writing, his "knowledge of the law was very small when I took him in (but) he would work hard and learn all there was in a case he had in hand… He would get a case and try to know all there was connected with it; and in that way before he left this country (the partnership) he got to be quite a formidable lawyer."

Lincoln formed a partnership with a bright young lawyer named William Henry Herndon, whose family he had known for a decade, and had even shared a bedroom with him above Speed's store. Their partnership, Lincoln and Herndon, was an arrangement that would last until his assassination.

Life on the legal circuit was hard. Lawyers traveled on horseback or in buggies over dusty and muddy roads no matter the weather. They rode through bitterly cold springs and sweltering autumns; they rode through storms and droughts, through snow and mud. They stayed in inns, boardinghouses and farmhouses, when necessary sleeping two or three in a bed, twenty men to a room and eating poor food. The hours were long, the pay was poor and even the beds were too short for his long limbs. Lincoln loved it.

On a horse, or later in a rickety carriage, he carried a few essentials—among them always some books—in a green carpetbag. Dressed in a shoddy dark suit and vest, warmed by a cloak or shawl, papers stuffed in his hat and carrying a well-worn green cotton umbrella with a broken handle that was tied closed with a piece of string, Lincoln rode the circuit for more than two decades. There were some who rumored that he preferred the deprivations of the road to the arguments of his home, but years later he spoke fondly of the camaraderie. The men were bound together by the shared hardships and formed strong bonds. They worked together on some cases and opposed each other on others. But at night Lincoln and these friends gathered around the fire, often in the judge's room, and shared stories and laughter.

Lincoln became known for his quick—and dry—wit. A popular story was told that one of their number was known for his exuberant gestures during his final arguments, causing his coattails to fly apart and expose his underwear. The men decided to take up a collection to repair the man's torn pants, but when Lincoln was asked for his contribution he explained, "I can't contribute anything to the end in view!"

It was while riding the circuit that Abraham Lincoln sharpened the skills that earned him a reputation as one of Illinois's finest lawyers. Meeting the challenges of the circuit was an ongoing adventure; they were called upon to handle every type of civil suit as well as criminal trials. With so many cases on the docket, lawyers had little time to truly prepare properly. Yet, a thorough understanding of the crooks and crannies of the law was an absolute necessity. Unable to carry many law books with them and having no access to a library, they had to rely on their own knowledge of the essential statutes. While Billy Herndon wrote that Lincoln was "strikingly deficient in the technical rules of the law," he added that his understanding of human nature and his insistence on doing the right thing made up for it. Lincoln might not have been able to cite the statutes verbatim, but he understood the spirit of the law; he knew that virtue was supposed to be rewarded and wrongdoing should be punished, and believed it was his job to make sure that happened. As Herndon continued, "He had a keen sense of justice, and struggled for it, throwing aside forms, methods and rules, until it appeared pure as a ray of light flashing through a fog bank."

Members of the traveling bar often had to try cases without even knowing all the facts. They were forced to put witnesses on the stand before uncovering everything that their testimony might reveal. Lincoln became a renowned examiner, able to draw what he needed from witnesses without allowing them to color his case. But above all he learned how to connect with jurors, presenting his evidence in an orderly, utterly logical and easy to understand fashion, and his summaries were said to be the best show in town. "In addressing a jury," reported the Danville *Illinois Citizen*, "there is no false glitter, no sickly sentimentalism to be discovered… Seizing upon the minutest points, he weaves them into his argument with an ingenuity really astonishing… He forces conviction upon the mind and, by his clearness and conciseness, stamps it there, not to be erased."

He was, said Isaac Arnold, who rode the circuit with him, "The strongest jury lawyer we ever had in Illinois… He could compel a witness to tell the truth when he meant to lie. He could make a jury laugh, and generally, weep, at his pleasure… He understood, almost intuitively, the jury, witnesses, parties and judges and how to best address, convince and influence them." Like every great lawyer, over time he developed his own courtroom manner. He rarely overplayed his hand, laying out the facts that supported his version of the affair. "Make few statements," Lincoln once explained, "for if I made too many the opposite side might make me prove them."

When his opponent presented his case, he mostly sat quietly and listened. He wasn't much for objecting.

Herndon noted that when other lawyers might have objected, he "reckoned" it would be right to allow the other fella to have his say. Those times he did object, if he was overruled he would admit with some chagrin, "Well, I reckon I must be wrong." But that was just strategy; most cases, he knew, turned on one significant point. He willingly conceded those points he couldn't win or that made little difference, building up goodwill with the judge and jury that might pay off when he pounced on the salient issue.

His ability to look at a case and whittle it down to those legal points that would carry the day was legendary. While he often spoke in a folksy way, and liked to include easily grasped comparisons, his presentations and arguments were lean and, when necessary, mean. His friend Judge David Davis, who presided over the circuit from 1849 to 1862, said, "When he believed his client was oppressed…he was hurtful in denunciation. When he attacked meanness, fraud or vice, he was powerful, merciless in his castigation."

It was this combination of qualities, according to the Danville *Citizen*, "that place Mr. L. at the head of the profession in this State."

His folksy manner was not always appreciated by the often self-reverential sophisticated gentlemen of the bar. Edwin M. Stanton, who later would serve in Lincoln's Civil War cabinet, once derided him as "A long, lank creature from Illinois, wearing a dirty linen duster for a coat and the back of which perspiration had splotched wide stains that resembled a map of the continent."

The nation had a great need for lawyers, who were being relied upon to determine how existing statutes might be applied to rapidly changing circumstances. They were shaping the future, determining how the rule of law should encourage progress. There were few precedents on which to rely. For example, few negligence attorneys are aware they owe a direct debt to Lincoln. Paved streets were still somewhat new in the west, so when a friend of his was injured after falling on an unrepaired Springfield street, Lincoln brought an action against the city, using the then-unique argument that under its charter the city had a legal duty to maintain safe streets. The opinion of the state's chief justice, Walter Scates, became a foundation of municipal law. "The obligation is perfect," he wrote, turning Lincoln's theory into settled law.

When rights conflicted, it was the job of lawyers to help sort them out. Few situations were more important to the future of the country than finding a way for the expanding railroads to coexist with economically vital riverboat traffic. For two centuries the nation had depended on its rivers to transport goods, but by the 1850s the railroads needed to build bridges to span those rivers. The riverboat folks were fully aware the railroads could cut deeply into the profits. Their attempt to stop them failed, and the first railroad bridge built over the mighty Mississippi connected Rock Island, Illinois, to Davenport, Iowa. A month after the bridge opened in April 1856, the side-wheeler *Effie Afton* struck one of the bridge piers and burst into flame, destroying both the ship and the bridge. The owners of the ship sued the bridge company, claiming its structure had unnecessarily created a deadly hazard on the water.

This was a fight to control navigation rights on the river. The case, which was tried in federal court in Chicago (and which Hitt transcribed), attracted national attention. As part of the defense team, Lincoln argued that the accident had been caused by the negligence of the *Effie Afton*'s crew. Riverboat pilot error! To prove it, he conducted his own investigation by chartering a steamboat and hiring experienced riverboat pilots to take him back and forth through the area at different times in different currents, until he was completely familiar with the vagaries of that stretch of the river.

The jury deadlocked, although reportedly favoring the defense. The judge dismissed the case, which was never retried, and the result was considered vital as railroads spread across the continent.

Lincoln and his colleagues literally were making up the law; they were establishing the precedents that future courts would come to rely on. An 1839 case, for example, in which a farmer who was accused of stealing a pig countersued his accuser for slander, helped reinforce the concept that truth was a complete defense against such claims. That same year Lincoln represented a traveling theater group that influential church members had tried to tax out of existence, causing the exorbitant tax to be rescinded.

Throughout the 1850s railroads provided much of his work, and he argued both for and against their rights and responsibilities. Working with Herndon, he represented a passenger who had been "pounded in the face ten or a dozen licks" by a conductor, Lincoln established the concept that companies were liable for the actions of their employees while working. Conversely, he successfully defended railroads against aggressive localities pressing tax claims and against stockholder claims, and even helped set an important economic precedent when he convinced the court that railroads should not be held liable for freight losses beyond their control.

He handled real estate cases, patent claims, he defended both men accused of seduction and women who had been slighted, and even represented nine temperance crusaders who in their zealotry had destroyed a "doggery," a low-class saloon. He actually lost that case, but successfully reduced the fine to $2 from each of the participants—which Lincoln supposedly paid himself.

He won many more cases than he lost, and in the process helped create an ethical standard for the law, rarely publicly criticizing a judge. Their power, he once explained, gave them "the last guess." But privately, when asked by his co-counsel how to respond after the decision had gone against them, he suggested, "Well, we can go over to the tavern and, just among ourselves, cuss the judge to our hearts content."

Lincoln was never so glib when discussing a criminal case. After all, freedom and lives were at stake. He defended men and occasionally women accused of illegal gambling, various forms of larceny and swindling, mail theft, running a house of ill repute, forgery, an entire family charged with assaulting a dinner guest, rape, attempted murder and murder.

It was an extraordinary array of cases. In 1841, for example, an eccentric old character named Archibald Fisher had disappeared soon after being seen in the company of the three Trailor brothers. It was rumored he was carrying a bag of gold coins when he vanished, which were to be used for a land purchase.

About a week later the postmaster received an anonymous letter claiming Fisher was dead and had willed William Trailor all his money, about $1500. The letter was made public and aroused great anger among Springfield's thirty-five hundred citizens. A thorough search was made of all the logical places, every basement and pit; recently buried horses and dogs were disinterred, but Fisher's body was not found. Henry Trailor was taken into custody and after three days of relentless questioning admitted that his two brothers, Archibald and William, had killed the old man and dumped his body in a pond. He told the story in considerable detail. His two brothers were arrested and charged with murder.

This was to become, wrote Judge James Matheny

a half century later, "Probably the most remarkable trial that ever took place in Springfield, and beyond doubt one of the most dramatic trials that ever took place in the whole country." Lincoln, Logan and Edward D. Baker, a renowned defense lawyer, were retained. Public opinion weighed heavily against the brothers, and there was real fear that vigilantes might organize a lynching party.

The prosecution presented what appeared to be an airtight case. In addition to Henry Trailor's confession, a respectable lady testified she had seen Fisher and the Trailor brothers enter the timber north of the town, and other witnesses swore they had found evidence of a struggle in the thicket and that the brothers had been spending gold coins. The circumstantial evidence was overwhelming; all that was missing was Fisher's body.

And then Lincoln for the defense explained the most serious problem with the prosecutor's theory: Archibald Fisher was quite alive. A day later the enfeebled man was brought into the courtroom. Years later Lincoln wrote about this strange case, explaining Fisher "had wandered away in mental derangement," and then pointing out that if he had died and his body was not found, innocent men would have paid a dear price. He added that several townspeople were greatly upset when Fisher appeared, quoting the "drayman Hart" who "said it was too damned bad to have so much trouble, and no hanging after all." Not all of Lincoln's murder trials ended as successfully;

to his dismay one of them did conclude with his client dangling on a scaffold. In 1838 twenty-one-year-old William Fielding Fraim, a cheery young Irishman, was working as a deckhand on a steamboat. He had been enjoying the spirits one afternoon when he got into a squabble with a ferryman named Neithamer, warning him to stop blowing "segar smoke" in his face. During that set-to he raised a butcher's knife and drove it deep into Neithamer's breast. The man was dead before his body hit the ground. This was Lincoln's second murder trial. He raced from Springfield to Carthage, Illinois, a distance of 115 miles, in only two days. Joining the proceedings in progress there was little he could do for Fraim; his claim that the indictment was "informal," meaning it was improperly worded and therefore should be dismissed, a strategy that had worked for him in his first murder trial, was denied. Fraim was convicted after a one-day trial and sentenced to hang. Lincoln's several appeals failed and, as a witness recounted, "On the day of the execution…the little town was thronged by men, women and children, afoot, on horseback and in wagons. Some came fifty miles, a few even a hundred, to witness the gruesome sight. School dismissed for the day." Many of the townspeople brought a picnic dinner with them for the event. There is no record that Lincoln attended, but it is clear he never forgot this failure.

Murder, of course, was the most sensational of all crimes, especially when it involved affairs of the heart. A good murder trial could provide entertainment for an entire town for weeks, and whatever the outcome the debate about it might last years. In 1856 the body of Springfield blacksmith George Anderson, whose skull had been shattered by one mighty blow, was found behind his house on Monroe Street. When his much younger widow stopped crying, she and the victim's nephew were charged with the crime. There were rumors of an illicit relationship between them and that the killing took place only after she had failed to poison him with strychnine.

The "crime of the century" riveted Springfield: young lovers conspiring to kill her husband so they might be together with his money. The newspapers carried two, three, four pages of stories every day, revealing every detail. And there seemed to be compelling evidence against them: the coroner had found traces of strychnine in Anderson's body—and a bottle of strychnine was found among the nephew's possessions. Servants revealed that George and Jane Anderson had quarreled the night of his murder, and he was so fearful that when he was last seen going outside to the privy he carried a gun with him. And in the nephew's trunk investigators had also found a flattering likeness of his aunt Jane as well as unsigned love letters!

Lincoln was offered $200 by concerned citizens of the city to assist in the prosecution, but instead accepted a quarter of that amount to represent the accused widow. He was joined in the defense by his former partners Stuart and Logan, and equally respected lawyers assisted the prosecution, leading the *St. Louis Republican* to report, "Greater amount of legal talent was exhibited in the argument of this case than in any case, probably, ever tried in this place before."

The defense parried every argument with its own theory. Others had eaten the same food as the victim so it could not have been poisoned. George Anderson was known to fear another man who had learned that Anderson had received some funds. There was no actual evidence of shared passion. The trial lasted ten days. The jury retired for only a brief amount of time before finding the defendants—not guilty.

No one else was ever tried or punished.

In the practice of the time lawyers could be hired by either the state or the accused, and Lincoln did prosecute at least one unusual murder case; one of the first known attempts in American legal history to use temporary insanity as a defense. The facts were indisputable. In the summer of 1855 Isaac Wyant and Anson Rusk had fought over a disputed tract of land. Rusk had shot Wyant in his left arm, which had to be amputated. Several months later the distraught Wyant responded, shooting Rusk four times in broad daylight. It seemed it was cold-blooded murder.

Lincoln prosecuted the case, although he had known Isaac Wyant for a considerable time. His close friend Leonard Swett, one of Illinois's finest lawyers, defended Wyant. It was a great show, two brilliant lawyers, friends, sitting at opposite tables in the courtroom. The recently coined word *psychiatry* had not yet entered the common vernacular, but Swett launched a spirited defense, calling a series of witnesses to create a portrait of an unbalanced man sent over the edge by the amputation. Lincoln countered as best he could, at one point responding to a physician's claim, "You say, Doctor, that this man picks his head, and by that you infer he is insane. Now, I sometimes pick my head, and those joking fellows at Springfield tell me there may be a living, moving cause for it, and that the trouble isn't at all on the inside! It's only a case for fine-tooth combs."

The jury set an important standard when it found Wyant "not guilty by reason of insanity" and "unsafe to be at large." He was committed to the Illinois State Hospital for the Insane. After the conclusion of the trial, William Herndon told Lincoln that he knew Wyant well, having defended him "from almost every charge in the calendar of crimes." Herndon described him as "a weak brother," easily led by others. Considering that, Lincoln responded, "I acted on the theory he was possuming insanity, and now I fear I have been too severe and that the poor fellow might be insane after all. If he cannot realize the wrong of his crime, then I was wrong in aiding to punish him."

Lincoln's most celebrated murder case, his defense of Duff Armstrong in 1857, almost directly preceded his defense of Peachy Harrison and certainly helped elevate his national profile. Even given all the drama surrounding murder trials, rarely did a verdict turn so completely based on the revelation of a single piece of surprise evidence. The killing took place at a rowdy Saturday night picnic in late August. Twenty-four-year-old Duff Armstrong had been drinking with some friends. According to witnesses, a fight started and James Norris smacked James Metzker on his head with a length of wood, then Armstrong hit Metzker in his right eye with a homemade weapon, a chunk of metal wrapped in leather called a slungshot. Metzker was more grievously wounded than anyone suspected. He mounted his horse and rode home, falling off several times. Once home, he took to his bed and died three days later.

It was impossible to determine which blow had killed him, so both Armstrong and Norris were charged with the crime. Norris was tried first, and was quickly convicted and sentenced to eight years. There seemed to be little hope for Duff Armstrong—and then Abe Lincoln took his case.

Many years earlier, when Lincoln had lived in New Salem, he had been invited to share the stark cabin of Jack and Hannah Armstrong. They had treated him as family. Hannah had cooked his meals and mended his clothing, and Lincoln had rocked their infant son to sleep in his cradle. Now a widow, Hannah Armstrong appealed to Lincoln to save her son.

Local citizens were so enraged by the killing that the trial had to be moved to the Cass County Courthouse in Beardstown, Illinois. As the trial proceeded,

Lincoln sat placidly while prosecutors made their case. A spectator remembered that when an eyewitness to the event, Charles Allen (whose testimony had led to Norris's conviction), was telling his story, "Lincoln sat with his head thrown back, his steady gaze apparently fixed on one spot of the blank ceiling, entirely oblivious to what was happening about him, and without a single variation of feature or noticeable movement of any muscle of his face."

When the prosecutor finished with his witness, Lincoln began his cross-examination and led Allen through the testimony he had just provided, focusing in on the details: Did you see the fight? Exactly where were you standing? Would you describe this slungshot? And finally, what time did this take place? He asked Allen several times if it was possible that he had not seen the event as clearly as he recalled. Allen remained adamant.

Lincoln had set his trap, and slowly began closing it. Allen testified several times that he had been standing no more than 150 feet away and had seen everything clearly in the brightness of the full moon.

Lincoln then famously produced an 1857 almanac. He turned to the proper page, then asked Allen to read the description aloud. Rather than a full moon that night, there actually was little more than a quarter-moon. And rather than shining brightly overhead at eleven o'clock; the moon had disappeared. It was practically pitch-dark. It simply wasn't credible that he could have seen the events he described from that distance on a dark night. The prosecution's pivotal witness had been exposed as a liar.

After several people testified to seeing Metzker fall from his horse, a doctor admitted that such a fall could account for the fatal injury. It also could have been caused by Norris's blow. Lincoln then produced a witness who claimed he had made the slungshot and described it in detail; Lincoln then cut open the leather sack to show jurors the weapon was exactly as described. That small slug of metal hardly seemed capable of inflicting a fatal blow.

In summation, after presenting his case he made an emotional appeal to the jury, telling them of his debt to Duff Armstrong's widowed mother in such vivid terms that he was moved to tears. He had been "a poor friendless boy, and Armstrong's father had taken him into his house, fed and clothed him, and gave him a home."

When the jury left the court to deliberate, he put an arm around Hannah Armstrong's shoulders as he escorted her outside, telling her, "They'll clear him before dark."

After only an hour the jury acquitted Duff Armstrong of all charges. Lincoln refused to accept any payment for his work.

After twenty years, this self-taught lawyer had risen to the top of the legal profession. His courtroom appearances drew large crowds who hung on his words. And that reputation helped position him to make a run for the presidency. But before the Republican convention at which its nominee for the 1860 election would be determined, he would become embroiled in one last murder case, his twenty-seventh. It was a difficult case that might well determine the legal boundaries for self-defense. And like many of his previous cases, he had a personal involvement.

But this time the nation was watching.

CHAPTER THREE

By the time Robert Roberts Hitt made his way the three blocks from the Globe Tavern across the public square to the Sangamon County Courthouse on Sixth and Washington, a large crowd had already gathered outside. All expectations were that this trial was going to be a humdinger, and no one wanted to miss a word of it. As Hitt walked sprightly up the steps, in between the imposing Ionic columns signifying strength and fortitude, several men acknowledged his presence, causing him to blush slightly at his reflected celebrity. "That's Lincoln's steno man," he heard someone say but kept his gaze straight ahead. A professional man like himself could not permit such attention to distract from his task.

As he knew, there would not be much work for him today. The first day of any trial, and perhaps even into a second, would be given over to motions and the selection of the jury for which transcription was not requested. It was here the tone of the trial would be set; the judge, in this case Judge Edward Y. Rice, would let the participants know how tight a legal leash he intended to keep. Each judge had his own opinion as to how a trial should be conducted; some of them "locked the doors," meaning they adhered strictly to the letters of the law, while others "let chickens in to play," meaning almost anything within reason that helped reach a just verdict was acceptable.

Judge Rice had earned a reputation as strict but fair. He and Lincoln had ridden the circuit together, but unlike the popular Judge David Davis, he had most often excused himself from the late-night banter in front of tavern fires. Although Lincoln would never say so aloud, Hitt was pretty sure he was not especially pleased to have Judge Rice presiding in the case. The judge was a fierce Democrat and at times had crossed words with Lincoln over political issues. While those differences had never carried over into the courtroom, Judge Rice was not going to give Lincoln too much trail to wander. He was not one of those judges who appreciated good courtroom wit and just let lawyers have their say; he kept a tight rein on conduct before the bench. Lincoln clearly preferred a loose courtroom, where he might make the occasional friendly comment that allowed him to share a laugh or a smile with jurors and helped build his relationship with them. One time, for example, after opposing counsel, Judge Silas W. Robbins had responded to a point he was making by claiming, "If that is so then I will agree to eat this desk." Mr. Lincoln had replied, "Well, Judge, if you do eat that desk, I hope it will come out a brand new manufactured wagon!"

The laughter that had filled the courtroom surely reminded jurors that this Lincoln was a fine fellow, who enjoyed using his sharp wit to prick pompous balloons.

But Judge Rice didn't allow for much of that breezy banter in his courtroom. Especially from his political opponents. Mr. Lincoln was going to get no benefits that might assist his political fortunes in this venue. This would be a fair trial with no room for amusement.

Hitt climbed the steps to the second-floor courtroom. The large wooden double doors had been tied open. Lincoln was already inside when Hitt entered, standing in a knot of people by the jury box. He was dressed as Hitt had guessed he would be: a black frock coat, dark trousers slightly too short for his long bony legs and dirty black shoes; his tie was slung to a side and his untended hair completed his presentation. To all appearances he was a homespun man of no pretensions—which made Hitt smile at this very pretension. Perhaps years earlier Lincoln might accurately have been described that way, but he had become a greatly sophisticated man, equally able to speak at length about the great poets or national affairs as to spout the law. And his practice had made him financially quite comfortable, certainly able to dress to fit his station.

Included among that group around him were several members of the defense team, but also the easily recognizable prosecutor John M. Palmer. Palmer was described by contemporaries as "broad-shouldered (with) a massive, splendidly shaped head and a facial expression that carried conviction. He had a convincing voice, a personality of tremendous earnestness and sincerity." Like all of the men who had ridden the circuit together, Palmer and Lincoln often found themselves as legal adversaries. But there was a strong bond between them. Both men had been raised in Kentucky and come to Springfield to practice law and politics; many had even remarked that it seemed they came from the same stable. Palmer served a term in the state legislature as a Democrat, but later joined the new Republican Party. When Lincoln made his run for Congress in 1855, Palmer, then a Democrat, had worked against him but their respect for each other ran so deep that Lincoln forgave the earlier slight and was there when Palmer presided over the first Illinois Republican State Convention in 1856. And there had even been rumors that while there the two friends had discussed the possibilities that might be found in the 1860 presidential elections.

Few men were as close to Lincoln as the prosecutor in this case, John M. Palmer. Lincoln had great respect for Palmer, whose integrity caused him to change his political party affiliation five different times to match his own beliefs. At their final meeting in 1865, Lincoln shaved while then General Palmer watched. He told the general, "You are home folks, and I must shave. I cannot do it before senators and representatives...but I thought I could do it with you."

Many years later Palmer would recall being with President Lincoln in the White House and admitting, "Mr. Lincoln, if I had known in Chicago that this great rebellion was to occur, I would not have consented to go to a one-horse town like Springfield and take a one-horse lawyer and make him President."

There were several other men in the group whom Hitt did not recognize; one of them he assumed was

Illinois's relatively inexperienced state attorney, J. B. White. In a murder case two years earlier, White was unable to prevent Lincoln from successfully postponing the trial for two consecutive court terms; Lincoln then won his client's freedom by asserting that the accused man had been denied his constitutional right to a speedy trial. There was tremendous public outrage against his client, who was accused of firing a shotgun into a crowd from inside his dark house, killing a popular husband and father. Lincoln had been granted the delays by asking for additional time to gather reluctant witnesses and requesting a change of venue until public anger quieted. By the time White was onto him, it was too late and the charges against the accused were dismissed. One could surmise that the humiliated White would be quite pleased to put a large rock in Lincoln's path right here, regaining some of his own lost favor. Hitt took his seat in front of the judge's bench, where he could easily hear every word. As he waited for the proceedings to begin, spectators filled all the seats and standing room in the back and on the sides of the courtroom. There were few women among them, but Hitt noticed with some amusement that the seated men tried to appear natural as they desperately avoided meeting the eyes of a woman, lest he would be compelled to give up his precious seat. But in several instances they were unsuccessful and, with a defeated shrug, gave up their seats and joined those standing.

While Lincoln's reputation was outgrowing the parochial courthouse, it was clear he felt quite at home

ABRAHAM LINCOLN PRESIDENTIAL LIBRARY AND MUSEUM

Springfield's statehouse cost $253,000 by the time it was completed in 1853. It was here that Lincoln first debated Stephen Douglas, gave his magnificent "house divided against itself" speech and eventually would lie in state. Many considered this the most magnificent capitol building in the west—although it would be replaced as the state capitol just a decade after Lincoln's death.

there, with good reason: even he couldn't calculate the number of cases he had tried in the county. He'd started his career in the old courthouse, which had been torn down in '37 to make way for the statehouse. In fact, he was one of the 101 good citizens of Springfield who had signed a note guaranteeing payment for the statehouse construction. The court had then been moved from place to place for almost a decade, causing more than one frustrated attorney to complain it "was movin' quicker than a man trying to keep his pay from his angry wife."

The court convened in a rented storefront for sev-

ABRAHAM LINCOLN PRESIDENTIAL LIBRARY AND MUSEUM.

Life in Springfield revolved around the public square. Here is an 1859 photograph of the east side of the square, taken by Preston Butler. Note the dirt streets of the capital, as well as the law offices of John McClernand and Norman Broadwell, both members of the prosecution team.

Broadwell later claimed it was his deft seating arrangement at the 1860 Republican convention that enabled Lincoln to win the nomination.

eral years, then for one year each in the Campbellite Church and then St. Paul's Episcopal Church, then back to the storefront until the new all-brick building was completed on the east side of the public square in 1846. County offices occupied the first floor, and a wide, sweeping stairway led to the large and simply appointed second-story courtroom. When designing the courtroom, planners had done as much as possible to prepare for the summer heat; it had particularly high ceilings so heat could rise above the spectators and included six large double-hung windows. Trees had been planted around the building to provide as much shade as possible, but they were still years of growth away from serving that purpose.

None of that had made the slightest difference to Lincoln, who had practiced his trade in one-room log cabins and large formal brick buildings. Lincoln had tried criminal and civil cases in all of Sangamon County's courtrooms, won some, lost some, settled quite a few, but noted with some pleasure that it was not the surroundings that made the law, but rather an adherence to the process.

The room had threatened to become overfilled by the time Judge Rice called the proceedings to order. There was some surprise that the trial actually was taking place so quickly after the crime. It was well-known that in almost every criminal case Lincoln tried to put space between the commission of the crime and the trial, letting time cool local passions and, if necessary, asking for a change of venue. There was little chance of that happening here but, with anger still at a boil, many wondered why he didn't at least make the motion. There was speculation that this was show, that he wanted people, including the victim's friends, to believe he was so confident in his defense that he was willing to try the case without delay.

The prosecution and defense had laid out their positions in the pretrial hearing weeks earlier. No one disputed the fact that Peachy Quinn Harrison had stabbed Greek Crafton during a brawl in Mr. Short's store. But the case would turn on why. *Why* was the all of it. Prosecutor Palmer intended to prove that this was a premeditated act, that Peachy Harrison had forced the issue with his taunts, then prepared for the inevitable meeting by borrowing the deadly weapon and carrying it on his person. This was not a random act of self-defense: Harrison hadn't picked up a knife lying on the counter to protect himself, he had been ready and waiting for the showdown.

The defense, conversely, was going to argue that Peachy Harrison had no choice but to defend himself with force, as he was by far the physically slighter of the men and Crafton had repeatedly threatened to stomp his face. During the confrontation, he had clung desperately to a counter railing but had been yanked free, then attacked by both Greek and his brother John.

The law defining self-defense while clear by statute was still somewhat murky in practice. The statutes had been loosely developed based on British common law mostly to protect white landowners and, like much criminal law, were subject to local interpretation. Friendly juries sometimes used self-defense to justify the verdict they desired; it was flexible enough to be shaped to fit popular outcomes. In a case earlier that year, Illinois Court of Appeals Judge Sidney Breese wrote that jurors "could assume the responsibility of deciding, each juror for himself, what the law is." But basically it was most often interpreted to allow a man to use as much force as reasonably necessary to defend himself from an attacker. Harrison's defense team had to demonstrate he was in fear for his life, as Lincoln had done successfully in the Truett case.

But Lincoln was holding what he hoped would be the ace. In his deathbed declaration, Greek had apparently claimed responsibility for his own demise. "I brought it on myself," he supposedly had admitted. Yes, he actually forgave Peachy. And while the man who took that declaration happened to be the defendant's grandfather, he also was a widely respected man of the cloth. Lincoln was not concerned about the Reverend Cartwright's credibility, particularly after taking an oath to the Lord, even to save his grandson's life. The problem Lincoln faced was getting this testimony admitted at all.

There had been no difficulty about that during the coroner's inquest. During that hearing Cartwright had testified that Crafton had forgiven his assailant. On the other hand, Dr. J. L. Million, who had comforted Greek as best was possible, testified that during the final few days he had several conversations with his dying patient and rather than absolving Harrison, he had censured him.

The rules governing the coroner's hearing were quite different than those followed during a trial. This clearly was hearsay and typically was barred from being heard by a jury. There simply was no way of knowing for certain what actually was said, and such emotional testimony easily could influence a jury. With Judge Rice on the bench, Lincoln surely knew it was going to require some clever lawyering to potentially get Cartwright's critical testimony admitted into evidence.

Just moments before the judge gaveled the proceedings to order, the defendant was brought into the courtroom. The marshal removed his wrist shackles, and Harrison took his seat at the desk between Lincoln and Logan, looking to Hitt as frail as a leaf. Hitt suspected this arrangement had been carefully planned, to make Harrison look even smaller and more vulnerable to the jurors. Too small, too weak to have stood his ground against an angry, hulking foe or foes. Logan whispered something to him, causing Peachy to smile wanly, then turn around to acknowledge several people in the front row. One of them, Hitt was certain, based on the fuss around him, was the Reverend Cartwright, Harrison's grandfather.

Lincoln had withdrawn several sheets of paper from his tall hat, which sat upended on the table, and was busily arranging them in some sort of order in front of him.

At half past the hour of nine the court crier, T. S. Kidd, shouted, "Hear ye! Hear ye! Hear ye! Court is now in session. Judge Edward Rice presiding. God save the United States and this honorable court." Judge Rice greeted the participants, cautioned spectators that he would tolerate no outbursts and immediately began jury selection. In most instances filling a jury box was a wearisome task; it seemed people like the concept of a trial by jury, but they liked it best when they didn't have to serve on the jury. At times the court would be forced to send marshals into the streets to bring back whatever bystanders they could find, but that was not true in this case. Being selected for the jury guaranteed a seat for the entire proceedings, in addition to the $1.50 a day pay. Just about everybody in Springfield knew both Peachy and Greek; a lot of them had seen both of the young men grow up, and already held an opinion about each of them. Greek could be rough sometimes, but he was at heart a good boy, one of Springfield's own. Maybe he brawled from time to time, but being killed for doing what most men had done? That was a different story. But Peachy was well-liked, too; he had an affable smile and a pleasant way about him, and wasn't free of shenanigans either. He wasn't a fighter like Greek, but wasn't known for being cowardly either. A lot of people wondered what had got into him to make him do something so foolish.

The difficulty in finding a truly impartial jury in this too common situation was stated accurately by the young humorist Mark Twain, who said, "We have a…jury system which is superior to any in the world, and its efficiency is only marred by the difficulty of finding twelve men every day who don't know anything and can't read." The law in Illinois made that even more difficult by specifying that jurors had to be natural-born or naturalized white, male, property-owning American citizens between the ages of twenty-one and sixty.

Lincoln, and Logan and Palmer, too, took great

care when selecting a jury. In an earlier murder case Lincoln had spent three days picking the jury and only one day to make his case in front of them, and got his acquittal.

His intention was well-known and simple. He would build a personal relationship between himself and the jury, a relationship strong enough that it required a mountain of evidence to overcome his claim that his client was innocent. This was his gift: by the end of a trial, the jury just didn't want to let him down. While most often the jury selection process was repetitive and dull, Hitt actually was looking forward to it. It was here Lincoln began building his rapport with the jurors. Lincoln's close friend Leonard Swett liked to tell people about the day Lincoln was questioning potential jurors for a circuit case in which they were serving as opposing counsel. "He asked each of them if they were acquainted with my client. Small town like that, everybody knows everybody else, it hardly was a necessary question. Of course they did, they said. Finally Judge Davis had had enough of that. 'Now, Mr. Lincoln,' he said, 'you are wasting the court's time. The mere fact that a juror knows your opponent does not disqualify him.'

"Lincoln looked at Judge Davis then explained, 'No, it sure doesn't, your Honor. But I'm afraid some of these gentlemen may *not* know him, which I reckon would place me at a disadvantage.'"

It is said that every great defense lawyer has a feel for those jurors more likely to be sympathetic to his case, an intuition developed over a long period of time and considerable experience. Lincoln was no different. He had his own way of determining who might be in tune with him, and sometimes it was just physical: he did not want blond, blue-eyed men on his jury, believing that type of man to be a bit nervous and too easily led by the prosecution in violent cases. Over time he had formed a prejudice against men with a high forehead, finding that too often men like this formed an early opinion and were hard to dissuade—unless he could discern from questioning that this man's opinion favored his client. He actually liked to have corpulent men on his jury, perhaps because it was generally believed that men of great size prospered by their wits rather than their toil. That was the kind of man who might better appreciate the delicacy of his arguments and approach, which Lincoln's partner Herndon described as the "patient but relentless unfolding of a case…reasoning through logic, analogy and comparison." Finally, whenever possible he wanted young men on his juries rather than older men. This was often because his clients were of a similar young age or experience, and undoubtedly he felt young jurors might relate better to the circumstances. And older men, he found, men like himself and Douglas and Stuart, were often too set in their ways and their beliefs to be moved. The panel had been summoned from the city and close farms.

It was considered an unfair burden to make those people living way out in the county responsible for fulfilling their jury duty. The progression of candidates for the jury continued through the morning. The different lawyers took turns questioning prospects: "Do you know the defendant or his family?"

"Did you know the decedent (or as the prosecution pointedly described Greek, 'the victim') or his family?"

"What is it you do to earn a living?" Once even, "Are you sober?" to which came the response, "You mean, right now?" At times the spectators would respond to an answer with a burst of laughter or even a supportive cheer. The first man seated was one of the oldest and more respected citizens of the city, Charles D. Nukolls, who was well-known to all the participants. It was an entirely appropriate choice. Mr. Nukolls had settled in the area decades earlier and worked in the leather business. With his profits he purchased land, including the tract on the corner of Sixth and Washington. He paid $12 for that lot and, as the city filled in around it, sold that property to the county in 1845 for $1200. It was the precise plot on which the courthouse, in which he was sitting, was built. More recently Nukolls had studied medicine at McDowell College in St. Louis and graduated from that institution. His Republican politics also were well-known, which obviously pleased Lincoln, and yet there simply was no way either side could challenge his ability to judge fairly in this case.

The questions continued. "Do you consider yourself a political man?"

"You think the leanings of any of these people (indicating the lawyers) will influence your ability to be fair-minded?"

"Have you read about this case in the newspapers?"

The second man seated, Josephus Gatton, also had arrived in Springfield before Lincoln. Like Lincoln, he had come to Illinois from Kentucky—in fact, Mr. Palmer had cautioned him, "You're not gonna let those Kentucky roots influence you for Mr. Lincoln, are you?" Gatton had earned his keep as a farmer, his success allowing him to continue purchasing and cultivating additional wild land and building large houses on the lots until he was one of the wealthier men in the city. As with Mr. Nukolls, it would have been considered an affront to reject a man like this for any cause other than his own choice.

By the end of the morning two more men had been selected. One of them, Moses Pilcher, who was acknowledged to be the best carpenter in the city, had been a political ally of Lincoln for decades. In an 1840 letter to his partner Stuart, Lincoln had suggested that Pilcher "be enlisted in the (William Henry) Harrison cause as he was always a Whig and deserves attention." Palmer, by law lacking the same number of juror challenges as the defense, and perhaps appreciating his future need for a fine carpenter, let him be seated.

The fourth was the widower Isaac Payton, who was struggling to farm a small spread with his two young boys while doubling-up as a shipping clerk for the new Wabash railroad line. Neither side wanted to deprive him of the juror's pay, which was sure to help him clear some debts.

There were some candidates who were dismissed by the judge, among them a farmer named Johnson who claimed, "I believe it's dawn when my rooster wakes me, other than that don't believe much a anything I hear," and a blacksmith who warned that if he was put in the jury "not one a you would get your horses done for a week!"

As the afternoon heated up, the jury box started to be filled. Zenas Bramwell fit Lincoln's desire for a corpulent man, in addition to being the son of a Baptist minister known for speaking in both Greek and Hebrew and being the brother of the town clerk over in McLean County. Bob Cass, who had one of the larger farms in the county and was known to be a stern but honest man, was approved without challenge. Ben Brown, whose prize bull, George II, was highly valued for mating purposes, took his place alongside the curmudgeon George Robinson, another farmer, who had lost a thumb and a big toe in a shooting incident. William Patterson was an immigrant and naturalized citizen who had landed in America a decade earlier and joined his cousin on his farm, and still spoke with a hint of a Scottish brogue. Although Lincoln shared the questioning with Logan and Cullom, Hitt noticed that even when just sitting and listening he remained wholly involved. It was clear he was a casual man, even in court, and on occasion when he stretched backward his very large black shoes would suddenly appear from beneath his wooden desk. At different times everyone in the courtroom had to fight to stifle a yawn; even Hitt was forced to surrender more than once, but if Lincoln did so the steno man didn't catch him at it. Sometimes he was staring straight ahead, sometimes his hands were clasped behind his neck and his eyes seemed riveted on a point on the ceiling, sometimes he was leaning forward on his right elbow, his thumb beneath his chin and his index finger curving along his cheekbone but, it was clear, even in this most tedious aspect of the trial he was paying close attention.

When it was his turn to ask questions, he turned on his folksy charm, which somehow never seemed disingenuous. He appeared right interested in every answer, as if listening to pick up that one speck of information that he might later hook on to and use for a personal appeal to that individual.

Occasionally Hitt glanced around the courtroom. His previous steno work had been in Chicago courtrooms, which were considerably different in feel if not in format. The basic courtroom was laid out precisely the same; he'd read somewhere that this was based on the British justice system. The judge sat at the head of the courtroom on a raised platform, purportedly because as the representative of justice participants had to look up at him, but also because it gave him a clear view of the entire courtroom. The witness stand was next to the judge's bench. The lawyers, the petitioners, the accused and most of the court officials sat or stood in front of the judge's bench, separated from the gallery by a low rail called the bar; called that because it was likened to the great sandbar in the Thames River east of London harbor that, once crossed at high tides, made ships and their crews subject to British law. While those Chicago courtrooms generally were more ornate, often with scribed ceiling moldings, balustrade railings, solid brass spittoons and tiled floors as opposed to wide wooden planks, even on that first day he noted another significant difference in the gallery. In the city, spectators dressed more formally and included far more women; in Springfield most observers wore work clothes of some sort, and perhaps because they knew the participants seemed more personally involved, more apt to respond with some sound or gesture, as if watching theater. Pretty much after each man had been interviewed and either chosen or most often rejected, he would be welcomed back into the gallery with a spatter of humorous remarks, ranging from, "You're too damn ugly for them to have to look at that puss every day," to "That's alright, they know you need your wife to make decisions for you."

In addition, Hitt saw two black men sitting freely in the Springfield courtroom; he couldn't remember ever seeing a black face among the spectators in a Chicago courtroom.

The afternoon droned on. Farmer William Hinchey drew hearty laughs when he responded resolutely, after being asked by Mr. Cullom if he was aware of the details of the confrontation, "I don't know *nuthin'* about *nuthin'*!" And then he looked proudly at the gallery and grinned a wide, mostly toothless grin. The gallery cheered as one when both sides agreed to place him on the jury.

Farmer Jefferson Pierce was the eleventh man to be selected, following an interesting exchange with young assistant prosecutor Norman M. Broadwell about how to reconcile his religious beliefs with the dictates of the law. Initially he professed some difficulty having to judge the consequences of sinners, not wanting to trespass in the province of the Lord, but after trading biblical passages with Mr. Broadwell he admitted he could see his way clear to reach a decision.

After concluding his questioning, Broadwell looked at Lincoln and indicated with a pleasant smile and slight bow that it was his turn. Lincoln responded with an appreciative nod and an equally knowing smile. Later Hitt would learn the basis of those gestures between the men. Six years earlier Broadwell had trained in the law with Lincoln and Herndon, striking out on his own two years later. He had been opposed in his very first case, an accusation of slander, by Lincoln. The parties had settled for an apology. But clearly Lincoln had watched this clever exchange with some personal delight, and accepted Pierce without additional questions.

A grayish cloud of tobacco smoke was hanging over the gallery, motionless in the still summer air, by the time jury selection concluded early in the evening. The final selection was twenty-two-year-old farmer M. H. Pickrell, the 101st man to be questioned that day and the youngest member of the jury. There had been very few contentious moments, and at times Hitt was perplexed why one or another of the candidates was rejected. Pickrell seemed mostly an unformed vessel, polite, unopinionated, wanting to please; he seemed ready to be shaped by whichever of the lawyers proved a more able legal sculptor; hence there simply was no reason to reject him.

Judge Rice made a few remarks warning the jury not to discuss the case or their deliberations with anyone, "that means everybody else on the jury, but especially your wives," and was preparing to dismiss court for the day when Bob Cass raised his hand. He was feeling poorly, he said, and asked to be excused. His word carried sufficient weight for his request to be granted without discussion. While no law at the time mandated there be twelve jurors, and they could have proceeded with the remaining eleven, both sides agreed that a twelve-man panel was preferable, assuming they could agree on one other person. So everybody had to sit back down.

Most of the remaining candidates had scattered, so Lincoln and Palmer agreed to seat a bystander, James A. Brundage, who had been watching the proceedings. Brundage was known to both men; in fact, Lincoln and Herndon were currently representing him in the appeal of a civil judgment. Mr. Brundage had sued Mr. William Camp in Sangamon County Circuit Court for $300 over nonpayment for the sale of two mules. When that court ruled for his opponent, Mr. Brundage had hired the firm of Lincoln and Herndon to make his appeal to the Illinois Supreme Court; Lincoln had argued the case but had yet to receive a verdict. Palmer knew him to be an honorable man and agreed to let him be seated.

By this time everyone was ready to get on home. Judge Rice reminded everybody to be back in the courtroom at exactly nine o'clock, or at least as close as they could make it, then gaveled the day closed.

Hitt gathered his tools, they were much too costly to leave in the courtroom, and watched as Lincoln whispered a few additional words to the defendant, spoke briefly to Logan, then put on his tall hat and walked out the door. Alone.

But as it turned out, he was not going home.

CHAPTER FOUR

Lincoln revered the concept of the law, if not always its application. As a young lawyer in 1838, he had been invited to make a few remarks to the Young Men's Lyceum of Springfield concerning the importance of the law to this still new nation. He began, in a florid style that he would learn through the years to whittle down, by urging "every American (to) pledge his life, his property and his sacred honor..." to maintaining the law. "Let reverence for the laws be breathed by every American mother," he continued, "to the lisping babe that prattles on her lap—let it be taught in schools, in seminaries, and (in) colleges; let it be written in Primmers, spelling books and in Almanacs; let it be preached from the pulpit, proclaimed in legislative halls and enforced in the halls of justice. And, in short, let it become the political religion of the nation."

More than simply a bouquet of words, this statement served as his personal guide to the practice of law. And it was why after a long day in a sweltering courtroom he was returning to his office rather than going home. There was more preparation to be done if he were to feel entirely confident. As he walked in the twilight, he began at the very beginning, reviewing each of the elements that had brought him to this stroll on this night.

His calendar had been full in July. He had resolved several suits lodged against various railroads, usually on favorable terms. On the fourteenth he had joined Stephen Trigg Logan and several other men on a nine-day trip on the Illinois Central to Chicago to assess the value of some railroad property in connection with a case filed the previous February.

Lincoln savored the opportunity to spend time with Logan, his early mentor and longtime friend. The two of them had been on every possible side of various cases, with and against each other. Once even, Lincoln and Herndon represented the plaintiff in a civil assault and battery case, while Logan stood for the defendant. When the circuit judge scheduled to try the case failed to arrive, the trust between them was so complete that Logan agreed Lincoln should take the bench and preside. The eleven-man jury found unanimously for Logan's defendant. They made an odd physical pairing: Lincoln a tall, lanky man with an angular, craggy face, and impossibly long limbs, that sometimes made it appear that all his parts didn't move in unison, while Logan was described by a contemporary as, "small in stature and frail in constitution, but a piercing deep-set eye and a large cranial development, imparted a highly intellectual appearance to his almost infantile features."

It appears that at some point as the express clattered along the rails Logan revealed to Lincoln that he had been retained by the Harrison family to defend young Quinn Harrison against the charge of murder. Lincoln undoubtedly knew the sketch of the case, but was also aware of its complications: the Harrisons and the Craftons were two of the most prestigious families of the region and now connected by marriage and murder, the case was sure to be fraught with unseen hitches. With the additional involvement of the Reverend Cartwright as a key witness, it also seemed certain there would be some political consequences. So it was with some reluctance, knowing the precariousness of Lincoln's own political position, that Logan asked him to join the defense team.

There was little gain in it for Lincoln. His name was already being bandied about as a potential choice for the Republican nomination for president. He had recently received a letter from his friend Samuel Galloway, a Columbus, Ohio, attorney, informing him that at the recent state Republican convention Governor Salmon P. Chase's "ultra ideas" about abolishing slavery had rendered him unelectable; Galloway had cautiously suggested to the delegates that Abraham Lincoln of Illinois would make a fine alternative. Within days Lincoln would respond to Galloway with his appreciation for this "very complimentary, not to say flattering letter." And while he agreed that the respected Chase, who was known nationally as the "attorney-general for fugitives," might not be the most formidable candidate, as for himself, well, "I must say I do not think myself fit for the Presidency."

But that possibility was in the air and it was a time to be cautious. Becoming involved in a case already drawing this sort of attention made little sense. In addition, his legal roster was already overflowing. There were numerous cases from the railroads and other corporations in which he was involved; almost daily he was receiving requests from clubs and organizations for speaking engagements (and he was planning a fall speaking tour); there were endless organizational tasks to be done for the new Republican Party and several potential candidates had written to him for advice; and there were the usual problems at home with the increasingly difficult Mary and their rambunctious boys.

Additionally there was the Cartwright problem. The seventy-four-year-old Reverend Peter Cartwright at that time was perhaps even better known and more widely respected than Lincoln. "God's Plowman," as he called himself, had been riding the backwoods circuit for almost half a century, bringing Methodism to the frontier. That was not especially fertile territory for the Lord's word, and this ministry caused him to fight both man and nature for survival. There were times when Cartwright had to rely on the strength in his fists as well as in his words. But he was credited with bringing many thousands of people into the fold. His autobiography, *The Backwoods Preacher*, published only two years earlier, had brought him to national prominence, and his presence in this trial would attract even more attention. And Lincoln and Cartwright had been at odds for a long time. They had run against each other in elections twice, with each man winning once, and their political division ran deep. Cartwright was a Jacksonian Democrat; strongly against slavery and far more outspoken about it than Lincoln. He advocated moral persuasion as the means to emancipation rather than political steps, which he believed would weaken the union.

As a man who appreciated the difference between facts and faith, Lincoln feared Cartwright's potent mix of religion and politics, especially in a charismatic man like "Uncle Peter," whose words were said to weave spells over his listeners.

Logan was aware of all these snags, he said, and yet he needed Lincoln on this case. He would serve Logan as a welcome partner in the strategic planning and the conduct of the trial. And perhaps most important, when Cartwright took the stand, Lincoln would question him.

Lincoln did not reply. Instead he responded with the barest hint of a smile in appreciation of Logan's unexpected gambit. Logan continued, explaining that Cartwright's testimony was sure to be suspect since he would be telling his story to preserve his grandson's freedom, perhaps his life. The prosecution undoubtedly would attempt to blunt the impact of that testimony by reminding jurors of the stakes. The disputes between Lincoln and Cartwright were publicly known; and if "Honest Abe" was willing to stake the reputation he had spent his lifetime building on the veracity of Cartwright's testimony, in addition to the pastor's own character, few men would dare doubt the truthfulness of it.

Lincoln agreed to consider the proposal, but both men knew he would do it. Their friendship, and trust, was such that Logan would never have asked if he didn't feel it was necessary or, conversely, if he believed there might be some harm in there for Lincoln's prospects, but in either event Lincoln would never refuse him.

Additionally, Lincoln held considerable affection for the Harrison family. The two families had been tied loosely together by marriage decades earlier back in Augusta County, Virginia. During the Black Hawk War he had served with George M. Harrison, Peachy Harrison's older cousin, who later remembered, "We ground our coffee in the same tin cup with a hatchet handle—baked our bread on our ramrods around the same fire...slept in the same tent every night—

THE LATE LLOYD OSTENDORF ARTWORK IS FROM THE LINCOLN COLLECTION OF THE LATE PHIL WAGNER, SPRINGFIELD, IL. IMAGE COURTESY OF WWW.ABELINCOLN.COM: ABRAHAM LINCOLN COLLECTIBLES DRAWINGS BY FAMOUS ARTIST

The Black Hawk War, in which federal troops and state militia pursued a band of Native Americans through Illinois and Wisconsin, lasted from April through July in 1832. Lincoln volunteered and was elected captain of his Illinois militia company and, as seen here, spent much of his time serving with Peachy Harrison's cousin George M. Harrison.

traveled together by day and by night in search of the savage foe—and together scoured the tall grass on the battle ground of the skirmish near Gratiot's Grove in search of the slain." And the boy's father, Peyton Harrison, was a friend and a sometimes client.

Logan made the financial arrangements. By the time they returned to Springfield on the twenty-second it was set. Fees had never played an important role in Lincoln's criminal practice since his corporate practice had been lucrative. But before the trial began, he insisted on visiting his young client to hear the entirety of the story directly from him.

This was common practice for Lincoln. He was a familiar face around the circuit jails, which ranged in quality from a single leg shackle affixed to the second-story floor of a dry goods store to a more secure barred cell maintained by a paid jailer. He had met his clients to hear their stories from their own lips, both men and women, in conditions from squalid to splendorous. He had sat in dank cells and in plush living rooms to hear the same plea: help me. There were times he advised the client to let him negotiate a settlement or quietly withdrew from the case after this meeting, but most often if he accepted the case he believed he was representing an innocent person and even those times when he was not wholly convinced of his client's role, he was determined to provide the best possible defense and let the law have its say. There was a story told about him that one afternoon, after having this conversation with a man who wanted to hire him, he said, "Well, you have a pretty good case in technical law, but a pretty bad one in equity and justice. You'll have to get some other fellow to win this case for you. I couldn't do it. All the time while standing talking to the jury I'd be thinking, 'Lincoln, you're a liar,' and I believe I should forget myself and say it out loud."

Peachy Quinn Harrison was being held in a room above the courthouse that had been provisioned for this purpose. He had remained in hiding, with the assistance of Shelby Cullom, until his legal representation could be arranged and his safety assured. There was bad blood now between the families and the possibility of a Crafton striking back was a real one. Logan had arranged for Peachy's surrender, and he was locked in this converted storeroom. The family had agreed to pay for round-the-clock guards.

Lincoln and Logan were let into the room. It was a small loft, a garret, with the sloping roof-ceilings making it impossible for Lincoln to stand tall wearing his hat. It had a single bed, a chair and a rather large window leading out onto the roof. A gray, dull light came through the window, clouds covering the morning sun, fitting for the somber mood that filled the room. It would have been simple to climb out that window and escape over the rooftops, but Harrison had nowhere to go. He was known on sight by most of the people living there, had no funds of his own and had never been more than two counties away in his lifetime. As Lincoln entered, he noticed a single shackle embedded in the sidewall, lying untouched.

Peachy Harrison stood to greet them. Peachy brushed his hands on his buckskin trousers before taking Lincoln's hand in a polite manner.

Lincoln often had been involved in cases with young clients, but with few exceptions they had been civil disputes. His criminal cases involving younger defendants had most often been about brawls or sexual assault claims; murders most often were the end result of a longer chain of experiences. People grew into murder; they didn't simply happen.

This was different, of course. Looking at Harrison, Lincoln was deeply saddened by the episode. The whole matter was a waste, an utter and complete waste. Harrison was a small, wiry young man. Against the two larger Crafton brothers he would have had little chance. "Well, this is a sorry matter," Lincoln said.

Harrison stared at the floor, nodding.

"I've known your family since afore you were born," Lincoln continued. "Practically before your father was born."

"Grandpap told me that. He said you keep hard opinions, but that you and Mr. Logan were the best law people around here."

Lincoln chuckled. "Well, we've had our differences for sure, your grandfather and me. So where's that name come from, Peachy?"

The young man smiled for the first time. "Always had it, since I was a boy. It's just an old family name, Sir. There's been lots of Peachys among the Harrisons, both men and women. I even got an aunt called Peachy."

Just as he said that, clouds covering the sun must have parted, because the room suddenly brightened a bit. "You know I knew Greek Crafton," Lincoln said. "He studied the law in my office."

"I do, I think he told me. We were friends sometimes." He paused, sighed deeply, then added, "Till all this."

"Why don't you tell us what happened. Just let it out."

Harrison glanced at Logan, who nodded, then began telling his story. "Didn't mean for it to happen," he said. He raised his right hand. "Swear to God. Nobody did. But I guess it just couldn't be helped." As Lincoln had been told, it had started at the Fourth of July picnic. Family matters, trouble in the marriage of William Crafton and Eliza Harrison, rumors of abuse. Harrison had said some nasty things. Greek had responded. Honor had become pride, pride had escalated into threats; neither of the young men knew how to back away without being ridiculed. Friends carried messages between them, making things worse. Tears welled in Harrison's eyes as he continued, but he wiped them away. There had been problems before, some fights. "One on one I wasn't much afraid of him," he said, the first scent of bravado he had shown. "Maybe I couldn't lick him straight up, but I would do okay. But the two of them, that wasn't right." He had borrowed the knife, he continued, but he was going to have to give it back soon. No, he hadn't meant to use it. It was just to protect himself, to keep them away.

That morning he was already in the store when John Crafton walked in. He saw him, but he didn't think anything was going to come of it. It wasn't John who had been making those threats. But when Greek came in he knew there was going to be trouble. When Greek threw his arms around him, he had taken hold of the counter railing, figuring as long as he held on there wasn't much he could do to him. After that, well, things just started happening fast. He didn't even remember going for the knife, he said.

Logan asked, "You're pretty good with a knife, right?"

Harrison considered that, perhaps sensing the danger in that question and clearly not sure how to respond. "I grew up with them, if that's what you mean. We all did. You know, when I was out on the farm. But I never ever cut nobody ever. I sure didn't mean to cut Greek like that, to hurt him like that. I just wanted to keep him away from me."

Logan explained the problem. "That's what we're going to tell them. But you know there's a lot of angry people with the Craftons, and they aren't going to see it that way. They're going to point out you went and borrowed a knife and you were carrying it with you. They'll say this was a plan. And that you weren't no stranger to using it. And that Greek didn't bring a weapon with him, he wasn't armed."

"He said he was gonna stomp my face," Harrison complained.

"Who told you that?" Logan asked.

Peachy hesitated, scrunching his face as he struggled for an answer. "You know, everybody. It was around. Everybody was talking."

Logan led Harrison through the events, at one point

asking him to stand up and show them the stabbing exactly as it had happened. "I didn't mean to stick him," Peachy insisted, but took them through the whole fight.

Lincoln said very little, listening carefully. Lincoln rarely took notes. "Notes are a bother," he once said, "taking time to make and more to hunt them up afterward." Instead he relied on a seemingly infallible memory.

Logan had explained to Harrison what to expect in the Grand Jury hearing and warned him that undoubtedly he would be indicted for murder and there would be a trial "which we expect to win." He cautioned him to speak to absolutely no one about the charges, even his friends and family, reminding him that they might well be called upon to testify and even words spoken with caution might be colored.

Lincoln finally stirred, telling him, "Listen to Mr. Logan, Quinn. Answer all the questions directly. If you don't know the answer, or you aren't certain, say that. But say only the truth. It's been my experience that no man has a good enough memory to be a successful liar." Then he asked softly, "Are you scared?"

Harrison nodded, his eyes watering once again, "Yes, sir, I am. Really scared."

"Good, that's a truthful answer. You're right to be. This is a dangerous matter. There's no need for us to hide from that. But if we all do our job, this is going to turn out right for us."

The Grand Jury hearing had commenced on the

thirtieth of August. Logan and Lincoln represented Harrison, while Eighteenth Judicial Circuit Prosecutor James B. White was assisted by former congressman John A. McClernand, a strong Douglas supporter, and Norman M. Broadwell. As the *Journal* reported, "The case excited unusual interest and the Court House was densely crowded during the day." The spectators waited expectantly through the long morning, but to their chagrin the proceedings were halted early in the afternoon and everyone was sent home when two significant prosecution witnesses, John Crafton and Silas Livergood, could not be located.

Lincoln appeared quite riled by the decision, believing it was incumbent on both sides to be prepared to meet the court's schedule. That was a lesson he had learned well on the circuit. He questioned why his side should be penalized for the failure of the prosecution to fulfill its obligation.

When the Grand Jury hearing finally began the following day the prosecution set out to establish the facts of the crime. Their initial witnesses testified that Peachy Quinn Harrison had inflicted the fatal wounds. Next Dr. Million testified that he had treated Greek Crafton; he had tried to save his life, but he had been cut too deep and there was nothing that could be done. Logan and Lincoln rarely interrupted, allowing the men to tell their stories in their own fashion. This was typical of Lincoln's strategy; he was not one to constantly interrupt or badger opposing witnesses, preferring to let them have their say then later, if he believed it necessary, attacking the truth of it.

A good lawyer, Lincoln knew, would begin to make his case based on this hearing. He wanted to hear the whole of the prosecution's case laid out. He wanted to look at their witnesses and hear them tell their stories in their own way. He wanted to get the feel of them, and listen for any conflicts in testimony. Two people in the same place at the same time testified to hearing slightly different words. Words have meaning in a courtroom, they have nuance and nuance leads to different conclusions; which one was accurate? Maybe neither of them was? But if necessary the inconsistency might be a way to impugn their testimony.

So Lincoln sat, not quite impassively, listening to the flow. Let them talk, he believed. Once he advised a young lawyer, "It is good policy to never plead what you need not, lest you oblige yourself to prove what you cannot." So he let them talk, remembering their testimony. And maybe later he would remind them of their words, and challenge them to prove what they could not.

Additional prosecution witnesses provided the background, explaining that the two men had quarreled at the Fourth of July picnic "because Harrison had told his younger brother that Greek was not a fit associate." Apparently there was some mention of the bad marriage. There had been no fighting that day, but Harrison and Crafton began making threats. Crafton said he intended to whip Harrison on sight and in return Peachy warned, reported the *Journal*, that if "Greek ever laid hands on him he should defend himself by shooting him."

None of the prosecution's witnesses dared claim that Greek was free of malice when he walked into Short's drugstore that morning. And perhaps he might be guilty of crimes, too, if he had lived. But he hadn't lived; Harrison had been waiting for him to attack; he had been ready with a cutting knife and he had sliced him dead. And that was the prosecution case for murder. The defense responded with eight witnesses. Lincoln and Logan had no real hope of avoiding a trial. This wounded community needed a full trial to settle the matter. The anger was too red-hot. So they presented their case in an orderly but perhaps shallow manner. Their key witnesses were the Reverends Peter Cartwright and John Slater, who ministered to the residents of Pleasant Plains. Logan questioned them. Before Peter Cartwright testified, he and Lincoln shook hands and publicly exchanged a few pleasant words, as was necessary, but mostly avoided each other.

The rules governing this hearing differed substantially from a trial, so the prosecution made no effort to prevent Cartwright or Slater from testifying. The Reverend Slater, who also attended the dying Crafton, reinforced Cartwright's claim that Greek had claimed responsibility for the brawl and forgave Harrison. In response, the prosecution called Dr. J. L. Million, who said he had spent significant time with the victim who, he testified, not only "did not absolve Harrison from blame, but censured him." The second day of the hearing was given over to the lawyers for a summation of the evidence. Once again the courtroom was packed, according to the *Register*, "(M)any being attracted there to hear the able arguments of the counsel in the case."

Broadwell and Logan presented their case throughout the morning, McClernand and Lincoln spoke through the afternoon. "The most profound interest was manifested," the local papers reported. "The grounds both pro and con were strongly contested, and the testimony thoroughly sifted."

The two judges finally ruled that Harrison should be tried for murder, a decision that aroused even more passion. The sentence for murder was hanging. At most, his supporters had believed, he would be tried for manslaughter, the penalty for that being a prison sentence. The justices also decided that he might be released on a whopping $10,000 bail, an amount suggested by the defense and which his family put up. "The counsel for the accused," it was reported, "although very clear that he committed the homicide in self-defense, suggested this course to the court, lest his release might appear like too hasty a disposition of so serious a matter."

John Palmer, who was widely respected by all, was brought in to lead the prosecution at the trial. Palmer was an honest man who shared Lincoln's faith in the law and at a point in their legal careers each had lost a murder trial and seen their client hanged.

They shared antislavery beliefs although neither of them allowed themselves to be described as abolitionists. While both men wanted to see an end to it, they feared the consequences of abrupt abolition and instead favored a negotiated settlement. Most important, Palmer made a fair and formidable opponent for Lincoln and Logan. And a dangerous prosecutor for Peachy Quinn Harrison.

All of those events had brought Lincoln to this walk in the twilight on the first day of the trial in early September. The stakes had been laid down, the evidence presented at the inquest and then in front of the grand jury, the trial jury had been impaneled. There were strong arguments to be made for both sides. As he always did before beginning his trial work, Lincoln took the jumbled facts of the case and sorted through them, trying to put them in order and determine where to make his stand. An important aspect of that meant thoroughly assessing the case against him. He once said that he habitually studied the opposite side of every disputed question, of every law case, of every political issue, more exhaustively, if possible, than his own side. The result of that, he continued, had been that in all his long practice at the bar he had never once been surprised in court by the strength of his adversary's case—often finding it much weaker than he had feared.

Undoubtedly Palmer, too, had been turning over his case to discover its weakest points. He had been with and against Lincoln too many times to go into a courtroom less than fully prepared. There would be no surprises, no last-minute witnesses, no newly uncovered evidence. With the exception of his production of the almanac at the Armstrong trial, Lincoln was never one for that sort of trickery. It was going to be a battle of preparation and wits; the law in its best form.

Winning also mattered to John Palmer, who had political aspirations of his own; there was talk of a run for the Senate, which might be impacted by the outcome of the trial. So it was not surprising that, just as Lincoln did, rather than returning to his busy home and his ten children, he took refuge in his office to once again review his strategy.

Unlike Lincoln and Palmer, Robert Roberts Hitt did return to his bed and board, the Globe, for the evening. The somewhat quaint charms of Springfield did not entice him, at least not yet. Tomorrow would be his first day of scribing, and he needed to be ready for the task. It was not an easy job, catching all the dialogue of a trial. While judges and lawyers tended to speak loudly, many witnesses tended to mumble or lower their face and speak into their chest. Those reluctant witnesses were an occupational hazard, and he simply did the best he could, filling in from memory or logic what he missed in the moment. He was sitting by a cold fireplace after dinner, enjoying the one glass of sherry he allowed himself each day during a trial. There were two other boarders in the sitting room at the time, a man about his own age and an older man enjoying an especially aromatic pipe. The younger man was perusing the latest edition of the *Harper's Weekly* newspaper.

Suddenly he laughed aloud. Hitt and the older man both seemed a bit perturbed by this intrusion on their privacy, but said nothing. But when he laughed out loud again it became impossible to ignore him; the older gentleman removed the stem from his mouth and asked politely, but with a tinge of displeasure, "May I enquire what has caused that outburst?"

The young man covered his mouth, and responded in a voice that resonated with Southern charm. "Oh, I'm so sorry if I've disturbed you, but this..." he said, indicating the magazine. "It's a wonderful story about legal wit. Perhaps I might share a bit with you?"

The older man indicated with his pipe that he should go on. Hitt turned a bit in his chair, as he read aloud, "'It is the business of a lawyer to be ready-witted; and it may be that he whose wit is sharpened in daily encounters deserves little credit for readiness. Mister Curren used to tell a story of Lord Coleraine, in his time the best dressed man in England. Being one evening at the Opera, he noticed a man enter his box in boots and vexed at what he thought an unpardonable breech of decorum said to him, "Why did you not bring your horse with you into the box?"

"'"It is lucky for you, Sir," retorted the stranger, "that I did not bring my horsewhip; but I will pull your nose for your impertinence!" The two were immediately separated, but not before arranging a hostile meeting. Coleraine went to his brother, George, to ask for advice and assistance. "I acknowledge that I was the aggressor, but it was too bad to threaten to pull my nose. What should I do?"'"' The young man paused here to build the drama, then glanced at his audience before delivering the grand conclusion.

Then he concluded, "'"'Soap it well," came the brother's cool fraternal advice. "Then it will slip easily through his fingers!"'"' The young Southerner laughed again, this time joined by Hitt. Yet even as Hitt laughed he couldn't help but see the similarities to the Harrison affair, an insult to be settled with a brother's assistance, and feel sadness that it could not have been so easily resolved.

The older man seemed less amused, forcing a smile then saying, "Well, I'm a lawyer and personally I haven't experienced such great wit from my fellows. As a whole, I find them to be a rather serious group."

Hitt could not help but seize the opportunity, saying somewhat boldly, "Are you familiar with Abe Lincoln? His ready wit is quite renowned."

"Of course I know about Lincoln," the gentleman responded defensively. "He seems quite a bit about himself, though."

"Here," Hitt said, repeating a story he had been told. In a case in which his female client admitted her guilt but asked for leniency, Lincoln made an appeal to the jury, telling them about a preacher who proclaimed during his sermon that the Lord had never mentioned a perfect woman. "At that," Hitt continued,

"Lincoln said a woman in the rear of the church stood and shouted, 'Pastor, I know about a perfect woman, and I've heard tell of her every day for years!'

"The minister was taken aback and asked about this woman who was not mentioned in the Bible. To which his parishioner replied, 'My husband's first wife!'"

Even the older man laughed at that one. "Tell me," the Southerner asked Hitt, "do you know Mr. Lincoln?"

Hitt blushed. "I do," he admitted. "I've actually recorded his speeches in the debates."

"You're Hitt!" the man said. "I'd heard you were staying here."

"Robert Hitt," the steno man said, standing and offering his hand.

"James Thomas," the young man responded, also rising. The older man identified himself as Cyrus Wright. "Well then," Thomas said. "You must tell us about him." And so Hitt did.

CHAPTER FIVE

By 1859, the United States was rushing headlong into its golden future. Oregon was admitted to the Union as the thirty-third state. In June, the Comstock silver lode was discovered in Nevada and the great silver rush began. In July, Amherst beat Williams in the first intercollegiate baseball game, 73–32, and daring balloonists set a distance record that would stand till the end of the century, flying 809 miles from St. Louis, Missouri, to Henderson, New York. In early August, the first passenger elevator was patented by Otis Tuft and weeks later America's first oil well was drilled near Titusville, Pennsylvania, marking the beginning of the oil rush.

But the foundation of it all had been built on shaky ground. One issue, slavery, divided the country like no other. Morality and economy had become intertwined, and unraveling them without grievous consequences seemed an impossibility. The issue had been put off, patched and covered over in an assortment of complex decisions that served mostly to delay what had become inevitable. In the 1857 Dred Scott case, the Supreme Court had ruled that slaves, former slaves and their descendants could not become citizens and that Congress did not have the right to prohibit the expansion of slavery to the territories. In response the antislavery movement grew rapidly as abolitionists in the north began funding radicals like John Brown, who advocated an armed uprising.

In June 1858, in accepting the Republican nomination for the Senate from Illinois, Abraham Lincoln had burst onto the national political scene proclaiming, "A house divided against itself cannot stand. I believe this government cannot endure, permanently half slave and half free. I do not expect the Union to be dissolved—I do not expect the house to fall—but I do expect it will cease to be divided. It will become all one thing or all the other. Either the opponents of slavery will arrest the further spread of it, and place it where the public mind shall rest in the belief that it is in the course of ultimate extinction; or its advocates will push it forward, till it shall become alike lawful in all the States, old as well as new—North as well as South."

Americans were beginning to accept the reality that the young nation could not move forward until the question was finally and unequivocally resolved. While some in the Republican Party demanded the end of slavery, there were as many differing opinions as there were candidates about how this might be achieved; these positions ranged from outlawing it immediately, by force if necessary, to prohibiting its expansion into new states and territories and allowing it to eventually die out. Lincoln had tried skillfully to keep his personal beliefs out of his political pronouncements, hoping that there might be some way found to lawfully end this inhuman practice. In August and September 1859 he still was considered a dark horse candidate for the Republican nomination the following year. New York senator William H. Seward was the favorite, but Ohio governor Salmon Chase, Pennsylvania senator Simon Cameron and former Missouri congressman Edward Bates also had their supporters. For Lincoln, the best strategy was to allow the other candidates to remain the focus, letting them publicly debate the pressing problems, knowing that in this divided country any firm stand would alienate a substantial share of the voters. Meanwhile, Lincoln burnished his image as a man of the people, as a brilliant lawyer defending not simply his clients, but the virtues of the law itself.

In Springfield, Illinois, the first week of September, the problems of the world and the nation were set aside as one thing captured their attention: the trial of Peachy Quinn Harrison for the murder of Greek Crafton.

The wooden sidewalks of the city had been pounded through the early morning as a steady line of people made their way to the courthouse. An early rainstorm that had left puddles in the mud streets had passed, and the humidity had already risen to an uncomfortable level. Hitt was in his seat, his pens a-ready when Judge Rice pronounced the court into session close to 9:00 a.m. The session began with a discussion about whether the named witnesses on both sides should be allowed to remain in the courtroom during testimony. This Motion for Sequestration, in which potential witnesses are prevented from hearing testimony that might influence their own accounts, stretches back to the Old Testament, when two elders who had been spurned by the beautiful Susanna claimed they had seen her involved in an adulterous affair in her husband's garden. She was about to be convicted when Daniel interceded, asking the judges to "Separate them far from each other and I will examine them." He asked each of them under which type of tree they had seen the act of intimacy. "A mastic tree," said the first elder. "An evergreen oak," said the second—and so they were convicted of bearing false witness. From that time forward it was accepted that potential witnesses could be kept out until they testified. But in this instance it was agreed by both sides that one man, the Reverend Cartwright, would be permitted to remain, while the other witnesses were asked to leave.

The prosecution called its first witness, Dr. John L. Million, a graduate of McDowell Medical College who had been practicing in Pleasant Plains since 1851. Although just thirty-two years old, he had already established himself as an astute businessman and had purchased several tracts on Sixth Street, acquisitions that eventually would make him wealthy. He was a small, bearded man, with spectacles, but nattily appointed to the inclusion of a stickpin in his lapel. He seemed a bit flashy to Hitt, a bit too well put-together, and he expected the doctor's answers would be professional and perhaps overly dramatic. Hitt liked to try to guess how witnesses would respond, and Million looked like a man who enjoyed attention. John Palmer conducted his examination, standing in his place behind his desk: "Where do you reside, Doctor?" he asked, and the trial had officially begun.

"At Pleasant Plains in this county."

"Did you know the parties here, the defendant Quinn Harrison and Greek Crafton in his lifetime?"

"Yes, sir."

Hitt glanced at Lincoln, who sat with his hands clasped in front of him on his table, leaning forward. His expression was set solid, his eyes locked on the witness.

"Where did they reside?"

"In the vicinity of Pleasant Plains. Mr. Crafton probably a couple of miles. The other probably three-quarters of a mile." Mr. Palmer had been looking at a sheet of paper held in his hands. He put it down and got to the crux of it. "Is it true…is it…that Greek Crafton is dead?"

The doctor grimaced, then nodded.

"What do you know of the cause of his death?"

This was a question Dr. Million obviously had been well prepared to answer. The objective was to describe the crime in its bloody detail, to disgust the jury with the fury and brutality of the attack. "He died from a wound inflicted in his side, I believe between the eleventh and twelfth ribs on the left side." Dr. Million poked that place on his own body with his left hand. "That wound penetrated into the cavity of the abdomen. We could not ascertain positively what organs were injured, but we supposed from the course the knife took that the spleen and the stomach were penetrated."

"When were you called to see him?"

"Well, sir," Million said, drawing out the word *well*, "I saw him, I suppose, a minute or two after this accident or affray occurred. He was out in front of the drugstore where the difficulty took place. This drugstore was in Pleasant Plains and belonged at the time, I believe, to Short and Hart. When I first saw him, he was in rather a staggering condition, when I caught him. He called to me. I ran up to him and he reclined on me in a rather failing position. I don't recollect which side. He was then taken to my house by myself and one or two other gentlemen in Pleasant Plains. I suppose it is something like a hundred yards from the drugstore, probably not quite so far."

Well prepared indeed, Hitt thought.

Palmer took him through his paces. "Was an examination made of his condition?"

Million squared his shoulders in preparation for delivering his testimony. "His intestines," he said, sighing at the memory, "a portion of them were protruding, and I returned them to their place and dressed the wound, taking two or three stitches and applied some plaster over it." Several people in the gallery gasped at Million's report, but Hitt refused to show any response. In fact, far worse had regularly been heard in courtrooms. The rapid spread of the railroads across the country had resulted in grisly accidents which had given rise to numerous lawsuits. So many, in fact, that in the 1852 *Haring v. New York and Erie Rr. Co.*, the frustrated judge had decided, "A man who rushed headlong against a locomotive engine, without using the ordinary means of discovering his danger, cannot be said to exercise ordinary care."

After sufficient pause Palmer continued, "Did you make an examination of his person at that time?"

"No further than the wound, and I think the seventh and eighth rib. It looked like a knife had struck against a rib a very slight wound. I should suppose these wounds had been made by a sharp pointed knife."

Lincoln sat passively. The description of the weapon as a "sharp pointed knife" was surely meant to impact the jurors. The meaning was clear: this was a weapon meant for attacking, sharp and pointed, rather than for protection. But Lincoln did not object, did not say a word. As his fellow attorney, Hiram W. Beckwith pointed out, he "relied on his well-trained memory that recorded and indexed every passing detail."

Palmer persisted through the details, asking about the direction of the cut. "It was being between two ribs, it was difficult to say, only from symptoms which way the course was," said Million. "We imagined that the knife took a horizontal direction. It would be horizontal if he was standing erect..."

Hitt chanced a quick glance at the jury, most of them sitting expressionless but focused.

Dr. Million continued, "About an hour and a half after the injury he vomited a large quantity of blood and that could not, I suppose, have got into the stomach without the stomach had been cut. It might have cut the duodenum."

Du-o-de-num was a term Hitt did not know. Obviously a part of the body. He spelled it phonetically, hoping it might be found in the edition of Noah Webster's dictionary he carried with him for just such a purpose.

"When was that?" Palmer asked.

"Saturday morning the sixteenth of July of the present year, 1859. In Sangamon County."

"You stated it was soon after the blow. Why did you come to that conclusion?"

"I saw Greek Crafton that morning. I think...on

the street. From the app…" Someone in the gallery began coughing and Hitt lost the remainder of that word. As he had been taught, the professional does not make assumptions. In a courtroom every word has a consequence. He was certain the word was *appearance* but did not mark it down. "…of the wound was inflicted between seven and eight o'clock, not noticing the time particularly, and he died sir, about eight o'clock Monday night. I think the death was caused by the wound."

"Do you know anything about the circumstances of the infliction of the wound?"

"I do not of my own knowledge."

"Are you a practicing physician?"

"Yes, sir."

Palmer dismissed the witness, and Judge Rice cautioned him to speak with no one about his testimony and be prepared to return tomorrow.

It was not at all surprising that neither Lincoln nor Logan saw fit to object a single time during the doctor's lurid testimony. It was easy for jurors to imagine the doctor stuffing Crafton's intestines back into his body, and the victim vomiting blood. It was a violent scene, there was no disguising that. But that did not impact Lincoln's defense. Besides, the accepted custom of the day was to let each side have its say; any objections would generally be made to substance rather than form.

Having established that Crafton had been knifed to death, Palmer then called Silas Livergood, an eyewitness to the crime. Livergood was a friend of Greek and John Crafton and had come to the store that morning with John. Livergood was a simple-looking farm boy, with a brush of wheat-colored hair that he habitually swept out of his eyes. But that was something of an illusion; Livergood also came from a comfortable home and had been educated. He had grown up with all of the fellows and knew them well. Had the fight gone differently, had Harrison been the victim, Livergood himself might have been in the dock, charged as an accessory to the crime. He raised his right hand and swore the oath. He was a resident of Pleasant Plains, he testified, and he knew all of the participants in the event and that, "I saw them both on the sixteenth at Pleasant Plains… The cutting affair took place on that day…"

Palmer led him through his eyewitness account. "Well, when I first saw them at the drugstore, they were together. Greek Crafton had hold of Harrison around the arm. He was back of Harrison. Harrison had hold of the counter." Asked what was said during the melee, Livergood responded, "I didn't hear Mr. Crafton say anything. Mr. Harrison allowed that he didn't want to fight or wouldn't fight him, I can't tell exactly. He said he wouldn't fight or said he didn't want to fight him. I can't give his language."

Hitt noticed Lincoln briefly glance at Logan, and they exchanged small smiles.

Crafton did not reply, Livergood continued, then described their positions. "Well, Crafton had hold of him at the time they were at the counter and I think Mr. Short had hold of him when Harrison made that remark."

With a piece of white chalk Palmer drew a "rude diagram" of the store on the floor in front of his witness, then handed him a long wooden cane and asked him to point out to the jury where this took place. Several jurors in the second row stood to better see the diagram. Logan stood; Lincoln did not but rather leaned forward. "This is the north end and this is the south," Livergood said as he poked the floor. "Here is the entrance on the north side and now there is where they were standing. There was a counter on the east side of that."

For clarity, again as he had been taught, Hitt added that the counter was "(running north and south)."

Livergood pointed to the spot near the counter where Harrison and Crafton were fighting. "Harrison was holding onto the counter. Crafton had him around the arms."

"What was Short doing?" Palmer asked, building his crime scene.

"He was not doing much of anything. He had hold of them, trying to part them. Both of them, I think."

"Did you see the hands of either party?"

"I saw the hands of Harrison. I am not certain but of Crafton. Directly after that John Crafton interfered and took hold of Short."

From his seat behind the defense table, Stephen Logan suddenly interrupted and demanded of the witness, "State if anything was said." Hitt was a bit surprised at this, as it was not something he had seen in the Chicago courtrooms. The decorum of the courtroom had long been settled, and lawyers were not to speak out of turn. But Judge Rice said nothing in response. Hitt supposed it was a carryover from the rough-hewn circuit courts where the rules were of necessity relaxed to get to the core of the issue. No one objected to this intrusion.

The witness turned to face Logan and replied, "John Crafton made some remark that I cannot give." Meaning, Hitt assumed, that he did not remember the exact words. Using the cane, Livergood indicated that John Crafton "was about this part of the store." Once again, Hitt further described the location "back" to convey the complete response. "As soon as Mr. Short took hold he interfered and he made the remark that 'Greek could' or 'should whip him' or something to that effect. Then they retreated to the back part of the store. Here was another counter. They all went together to the back part of the store."

"Where were you standing?"

"About there, in the front part of the store. Harrison was nearest me as they were retreating… They retreated in front of this back counter," Livergood continued, pointing. "There were some boxes standing there and they retreated to the boxes and there Crafton got into a leaning position and the cutting was done there."

This testimony clearly was the crux of the matter.

Lincoln had propped his elbows on the table, and his palms formed a church in front of him as he listened intently. "I don't know what placed them there," Livergood continued. "I saw Harrison at the time. He was in the same position in which he was when he started, with his back to them. He had not changed his position. Harrison did the cutting. I saw it. Crafton was in a leaning position. He was leaning over the boxes half down." Livergood leaned over to demonstrate.

Hitt was not a lawyer, but he realized the danger of this testimony to Harrison. By Livergood's account, Peachy was in no obvious distress; he was pushed around but neither his life nor even his safety was threatened. Yet still, he had responded with deadly force.

Palmer had what he was going after, but continued, "What was the cause?"

Livergood shrugged. "I suppose retreating was the cause."

"Was the retreat voluntary or forced?"

"Forced, I suppose."

"Who applied it?"

"I suppose Mr. Short and John Crafton."

Palmer's voice seem to rise, his pace got a little quicker. "How were the blows struck?"

"Harrison was standing on the east of Crafton and he struck him right in the side. Crafton was inclined in this way." He turned to his side and glanced at the jury to ensure they understood. "Harrison stood right in front. I saw two blows struck with a knife."

Palmer hesitated to allow that information to sink in. Two blows. Not one. Not an accident. Two blows. And then he asked his important question. "What was Crafton doing when he was struck?"

Livergood answered in a voice tinged with anger and accusation, "He was not doing anything. He was in that inclined position. I saw the knife. It was a white-handled knife."

Logan interrupted once again, asking, "Have you detailed the facts fully up to this time?" And again, there was no complaint about it. Hitt wondered about the purpose of the question. Was Logan laying a foundation on which the defense would be built?

If it was intended to trap Livergood, it failed. "I don't know as I have," he said. "I don't believe I have. I don't know what I have omitted. If I have omitted anything, I don't know it." A perfectly ambiguous answer.

Palmer resumed. "Had anybody hold of Harrison when he struck the blow?"

"Yes, sir. Greek Crafton had hold of him by his left arm."

"Assuming that I was Harrison, how did he have hold of him?"

Livergood left the witness stand. Palmer turned around to directly face the jury. Livergood took a position behind him and swept his left around his body, pinning Palmer's left arm. "Crafton had hold of him something like that, with his arm around his arm."

Still looking directly at the jury, Palmer asked, "With which hand did he strike?"

"With his right hand. It was loose at the time."

"How long had that hand been loose?"

"Just then."

"Had you observed it before that?"

"Yes."

"Was anything in it?"

"No, sir. There was nothing in that hand."

"In what position was Crafton when the knife was drawn?"

"The same position he was when he was cut."

All this while Livergood retained his hold on Palmer. The visual evidence was strong; Harrison was in no mortal danger. Crafton was holding him; he was not choking him or stomping him or punching him. He was restraining him. It was no more threatening than might be seen a dozen times a day when children fought.

"Describe the knife," Palmer continued.

"I suppose the blade was about four inches long. A white-handled knife."

"How long did they remain there?"

"They did not remain there." Palmer removed Livergood's arm from around his body and separated. Livergood took the witness chair again and continued, "I could hardly tell how they were separated. I know I went toward them and helped separate them."

"What happened after the blows were struck?"

Even as he recorded these words, Hitt wondered

why Palmer had neglected to ask his witness to demonstrate how the two blows were struck. It seemed an odd omission.

"I don't know as Mr. Short did anything. If he did I didn't see him do anything. Greek Crafton got out of the house and directly after Greek was cut, John was cut. John was making toward Quinn when he was cut."

"Where was John when Greek was cut?"

"John was behind Quinn like. He was in a leaning position too. Mr. Short had hold of Greek and Quinn. He let them go immediately after the cutting was done. After Short let him go, Greek got out of the house and John commenced throwing a pair of scales at him (Harrison), and a stool. They were all loosened up when John was cut. When John received the blow he stood to the south of Harrison and Greek to the west of him… I think that was the position they were in at the time."

Palmer sat down at his table again. "What became of Greek Crafton?"

"After he got out of the house he was carried to Dr. Million's residence in Pleasant Plains and they commenced dressing his wounds. He lived from Saturday morning until Monday night. The night of the eighteenth he died at Pleasant Plains."

"You spoke of John throwing something. Was that before or after he was cut with the knife?"

Livergood answered firmly, "*After* he was cut with the knife." Palmer leaned back in his seat and allowed a few seconds to pass before nodding to Judge Rice and saying he had no more questions. The judge pointed to the defense table. Logan stood. He would begin the cross-examination. But in the relaxed atmosphere of this courtroom, it was accepted that Lincoln might also question the witness as he desired. There were no rules against this double-teaming, as it was known.

Logan took a position directly in front of the witness. There was a somewhat condescending edge to his tone as he asked, "Was there not a little circumstance you omitted? You say as they went back Greek had hold of Quinn with both hands?" Livergood acknowledged that was true. "You said when he drew his knife Greek had but one around him?" That too was true. "How long was it before Quinn struck did he take his hand away?"

"Immediately."

"What did Greek do when he took his right arm from around Quinn?"

"Greek struck him in the face." Hitt kept his head down so no one might see his smile. So that's what Logan was after. "Immediately after Greek struck him in the face he drew the knife."

Logan pressed his advantage. "Did Greek give him a pretty severe blow before he drew the knife?" That was it, that was key to it all. Several jurors in the first row leaned forward.

"I don't know," Livergood answered, either honestly or well prepared. There was no way of knowing. "I can't speak for that."

"Did you see the mark?" Logan asked, implying that Greek had struck such a strong blow as to leave a mark.

"No, sir."

"It was not until Greek keeping hold with his left and letting go with his right hand struck him in the face that he drew his knife?"

"No, sir, but immediately after that Quinn drew his knife." Logan was carefully countering the damage that Palmer had done to the defense, putting Peachy Quinn in real danger. "You say when you first went in you saw Greek had hold of Quinn around the body and arms with both his arms?"

"Yes. Quinn was holding onto the counter."

Logan looked at the jury as he asked this next question, to make sure they understood its importance. "The first thing you heard when you went in was by Quinn, 'I don't want to fight you and I won't fight you?'"

Palmer sat still as Livergood agreed that was correct. "Yes. Greek still held onto him and he held onto the counter."

He was drawing a strong image of Harrison desperately trying to protect himself. He said clearly he did not want this fight. He held onto the counter to prevent it. "What broke him loose?"

"Mr. Short and John Crafton interfering. That

pulled Short back. Short, I suppose, was trying to separate them and John interfered."

"To prevent their being separated?"

Livergood shook his head. "I don't know about that."

Logan continued, "John made the remark that, 'Greek could whip him and should whip him'?"

"Yes, sir. I don't say that that is exactly the words. When they retreated back the balance took Quinn back I suppose. He was dragged back from the counter."

This was the point Logan was trying to emphasize: Quinn did not want this fight. "Was he not dragged back with such force as to break his hold on the counter?"

Livergood shrugged as he answered, his reluctance visible. "Yes, I suppose so. I think he was holding onto the bar of the counter."

For the first time, Lincoln spoke. His voice, which at times tended to rise and even become strident, was soft and even. A curious friend asking a question. "Was there not an iron railing?"

"Some witnesses said so," Livergood acknowledged. But he couldn't be sure. "I took him to be holding onto the counter and trying to be kept from being pulled away."

Lincoln nodded, and stood. As he did, Hitt deftly dipped his pen twice into the inkwell. He had brought four pens with him, two gold and two steel. The gold nibs, while costing as much as $4, wrote easily although they tended to wear quickly. He had been using these gold pens for several months and was aware they would soon have to be replaced. He checked his open inkwell to make certain an ample supply remained. For this assignment he used a simple black ink purchased at a Chicago stationers.

Lincoln glanced down as if reading from nonexistent notes, then asked, "How long had Greek Crafton been in before you went in and saw him with his arms around Quinn?"

"Only a little while. I started from Turley's store with him."

"What was said when you started about what you were going for?"

Livergood answered strongly, "Nothing was said at the time."

Lincoln persisted. "Did you go in concert with Greek Crafton?"

"No, sir! He asked me to go along with him. He did not tell me what for."

"Objection, your honor!" Palmer stood, waving his hands. "This is not proper cross-examination."

It was here for the first time that Hitt got a sure glimpse of the underpinnings of the defense case. Lincoln intended to demonstrate that when Greek Crafton found out Harrison was at Short's store he set out to beat him. But he wasn't going alone, there were three of them: Greek and his brother who was waiting at Short's for them and Livergood. Lincoln told Judge Rice flatly that he was going to make his case that his client had good reason to be a-feared for his safety. He was firm in that, and said he would either show it now or later.

Palmer reminded the judge that his witness had been firm that was not at all the situation.

Judge Rice finally ruled that this was not the proper time, or witness, for Lincoln to prove that. "Bring it in on your own witnesses," he told him.

Having had his say, and confident the point had been made to the jury, Lincoln sat down and Stephen Logan resumed questioning the witness. Logan had sensed an advantage in Palmer's demonstration and decided to press it. He asked Mr. Broadwell, a member of the prosecution team, to assist him. He positioned Broadwell in front of the jury and took Crafton's place behind him, wrapping his arms around the prosecutor. "Is this the way they stood?" he asked Livergood.

"Yes, sir. And Quinn held his arms out to the counter." Logan directed Broadwell to grab hold of the jury railing. "When he was dragged loose they went in about the same position. When Greek let go with his right hand and struck him in the face with it, still holding him with his left hand."

Logan directed Broadwell to let loose his right arm and raise it. "How did Quinn's position change before he struck him?" he asked.

"He kind of turned around with his side to the Greek."

"What side of Greek were the wounds on?"

"The left side."

Logan deftly turned Broadwell to his left. "Then he was standing in about that position and had to strike around to hit Greek?"

"Yes."

"Was he so that he could see where he was striking?"

"Yes, sir."

Logan paused dramatically as if Livergood's testimony was confusing. "But it was in this position?"

"Yes, sir."

He let that confusion drip into his voice, wondering, "But he had to stoop around in that position and Greek still had hold of him?"

"Yes."

"And Short was holding him too, was he not?"

"Yes."

Logan put his hands on Broadwell's shoulders and gently pushed him into a stooping position, then turned him to the side. "And after receiving the blow in the face he stooped around in that way?"

It was an awkward position and made it appear as if Harrison was being overwhelmed. Livergood tried to limit this damage, telling Logan, "It was not a blow struck as if standing loose in this way. I suppose much more of his arm was loose." He got out of the witness chair and took hold of Broadwell in a manner that did not restrain him. "Short had hold of him so, and Greek had his left arm around Quinn's body." He let loose and stood tall. "And in that position Greek was struck."

The adjustments Livergood made were slight, but significant, making it seem as if Harrison was being restrained rather than held firm and unable to move.

"What was John doing then?" Logan asked, reminding the jury of the odds.

"Not much of anything. He had just been pulling back but he could not pull away further because he got against the counter. It was before that he said Greek should and could whip him." He mused, almost to himself, "The whole thing occupied very little time."

"So that not long after John said that the blow was struck, first by Greek and then by Quinn?"

"Yes."

"Did you say anything while it was going on?"

"No, sir, not a word."

Livergood spent the remainder of the morning on the witness stand. There were many more questions about the precise positioning of the participants and the timing of the events. Both Lincoln and Logan had their chance: Could Livergood recall John Crafton's exact words? He could not. Did Harrison say anything? He did not. Did he demand to be let loose? He struggled. Did he threaten Greek? He didn't say a word.

Palmer was quite satisfied with the morning's work. He had demonstrated to the jury the fury of the attack and had shown them how it happened. Nothing in the demonstrations, either his own or those staged by the defense, positioned Harrison in any life-threatening way. In fact, it seemed Mr. Short was trying to break up the fight when the sticking was done.

Lincoln and Logan also were satisfied. They had planted the seeds for their case. They had filled the room with the Crafton brothers and Livergood, all of them out to beat Harrison. While they had not gotten much of their own case in yet, as expected, they had limited the damage from an eyewitness.

Livergood finally stepped down off the witness stand. Hitt watched him walk past the lawyers' table to the gallery; as he passed Harrison he glanced at him, acknowledging him with a slight nod. When he got to the back of the courtroom, he was surrounded by a group of young men, who pounded his back and reassured him he had done right by Greek. Lincoln and Logan sat at their table, both of them facing Harrison and, Hitt guessed, bucking up his confidence. Hitt wondered for a moment what must be running through Harrison's mind. To be the cause of all this, to feel the breath of mortality at such a young age, and all because of a stupid action.

His reverie was interrupted by Judge Rice declaring a brief break in the proceedings. A "necessary break," Hitt assumed, as the judge banged his gavel and was gone into his chambers. The melody of courtroom noises erupted, replacing the respectful silence; spectators stood and stretched, stamped on the wood floor to get the blood flowing back into their legs and began discussing the testimony of the morning. In essence, the overture was done; the stage had been set, the players had made their introductions and now the drama of it could unfold.

The next witness to be called would be Greek's brother John Crafton, who had been at the center of it all.

CHAPTER SIX

Robert Roberts Hitt straightened his tools, then reached into his carpetbag to fetch his reading matter, in this case the May edition of *Ballou's Pictorial Drawing-Room Companion*. He had learned early in his career that a trial is a series of starts and stops, a precious few dramatic highlights separated by long periods of tedium and waiting. There were often long pauses, for lunch, for conferences, for personal needs, and so he always carried with him a book, magazines or newspapers to fill those spaces. Hitt opened the magazine, which had been published in Boston, and began leafing casually through it. He noted with satisfaction it contained several illustrations by the artist

Winslow Homer, whose work he thoroughly admired. As he did, all around him people found ways to fill the time. Several of them had gone outside to relieve themselves or have a smoke or chew, but many more had stayed in place, wary of giving up their seats. The room was filled with the harmony of numerous conversations, one of them rising loudly above the din for a few seconds as a point was made, then fading, replaced immediately by another one from a different part of the courtroom, all of it occasionally punctuated by bursts of laughter.

Hitt looked away from his magazine for a moment, leaned back in his chair and took all of it in with a great sense of appreciation: the sounds, the scene, the feeling of comfort he found in the predictability of an American courtroom. With the nation roiling over the insufferable question, with some people even brazenly discussing its continued existence, he marveled at the very normalcy of it all. As much as anything else, even more so than all the economic progress, this proceeding was the living result of what the founding fathers had fought for, a nation governed by laws.

The right to be tried by a jury of your peers, without the government having a say in the outcome, was one of the fundamental individual rights over which the Revolution had been fought. The first Continental Congress in 1774 warned that colonists were deprived of their basic rights as English citizens, among them "the great and inestimable privilege of being tried by peers of the vicinage, according to the course of that law." The Declaration of Independence blamed King George III for "depriving us in many cases, of the benefits of trial by jury." It had been so important to Jefferson and Adams and all of the other founders that it was guaranteed by the Constitution in several places, and detailed more often than any other right: Article III Section 2 provided the right to a trial by jury to anyone accused in a criminal case; the Fifth Amendment insured that no one might be arrested or tried for a "capital or otherwise infamous crime unless on a presentment or indictment of a Grand Jury"; the Sixth Amendment guaranteed that in a criminal trial "the accused shall enjoy the right to a speedy and public trial, by an impartial jury"; while the Seventh Amendment defined instances in which litigants in civil cases were entitled to a jury trial.

The rules by which Americans agreed to live together were laid out in that hallowed document. The acceptance of it had been so complete that most people now took those rights for granted. A good trial was considered grand entertainment, a welcome diversion from the chores of the day. While rarely did the outcome have an effect on anyone except the participants, a trial provided a welcome opportunity for spectacle and grist for hours of debate. A trial like this one, in which all of the participants, the victim, the accused, the prosecution and the defense lawyers were all a part of the town stew, allowed for even more personal involvement.

As Hitt glanced once more around the room, still

catching bits of conversation, he realized how few of these people were aware of how much and how long it had taken to make these proceedings seem so perfectly normal. The only reason he knew anything about that legacy was that when he had chosen his profession he had dug into the history of the courts system, his vibrant curiosity driving him to learn about its underpinnings. It had been a long and literally torturous journey to get to this day. Before the twelfth century the legal system depended primarily on Church-established ecclesiastical courts, which believed that God protected the innocent and so relied in criminal cases on "ordeal," a physical test like carrying a red-hot iron a certain distance without blistering, to determine guilt or innocence. Civil cases were resolved through "compurgation," in which the litigant producing the most witnesses willing to take an oath in support won the case.

It was British King Henry II who established a rudimentary jury system during his struggle for supremacy with the Roman Papacy, allowing common citizens to choose between trial by jury or combat to resolve disputes. Although in criminal cases these royal courts still relied on ordeals. The right to a trial by jury probably became the bedrock of the legal system in May 1215, when landowners, barons as they were known, forced King John at knifepoint to sign the Magna Carta on the meadow at Runnymede. One clause of it read, "(N)o freeman shall be taken or imprisoned or seized or exiled or in any way destroyed...

except by the lawful judgment of his peers and by the law of the land," guaranteeing forever this basic right.

In his research Hitt had found no agreement on when and where the very concept of a trial by a jury of a man's peers had originated. There was some reference to juries in ancient Athens as early as 400 BCE. These citizens did not apply the law, wrote Aristotle, but rather listened to the facts then reached a decision that met with their "understanding of general justice." The legendary English jurist Sir William Blackstone himself had written, "(W)e may find traces of juries in the laws of all those nations which adopted the feudal system, as in Germany, France and Italy, who had all of them a tribunal composed of twelve good men and true… Its use in this island…was always so highly esteemed and valued by the people that no conquest, no change of government, could ever abolish it… In Magna Carta it is more than once insisted on as the principle bulwark of our liberties…"

The concept that a jury vote must be unanimous for a conviction was accepted as early as 1367, but at different times in various places verdicts by majority vote had been permitted. In Springfield, Illinois, though, in 1859, the law was not fickle; a man's guilt must be proven so completely that a unanimous jury of his fellows had no reasonable doubt of it.

Hitt's reverie was interrupted by the court crier, Captain Kidd, refilling his water glass. Thomas Winfield Scott Kidd was a burly, balding man with slicked down gray sidewalls, a man who might be described as jolly looking; the hair missing from his pate appeared to have migrated down, flowing around his mouth to form a horseshoe mustache. He was dressed to add a degree of formality to his position: a dark jacket covered a sparkling white shirt topped with a light-colored bow tie. Captain Kidd smiled pleasantly as he poured the water. He recently had been appointed to the job by Judge Treat, after serving for seven years as a deputy under Sheriff J. B. Pirkins. In the tradition of town criers, his responsibility was to make the necessary announcements, including opening the session, calling witnesses and closing the day, all done in a deep baritone. While many court criers favored the traditional opening "Oyez! Oyez! Oyez!," a French term meaning literally "Hear ye," Captain Kidd shunned it because of its common use in British courts, and instead shouted the English words, "Hear ye! Hear ye! Hear ye!" three times for good luck, then adding the name of the presiding judge and concluding, "God save the United States and this honorable court."

"I trust you are enjoying our city," he said pleasantly to Hitt as they waited for court to resume.

Hitt nodded. "What I've seen of it."

Kidd raised his eyebrows in resignation. "Whatever it is you've seen, there isn't too much more to see. But if you have a minute you might wander over to the stables," he continued, tilting his head toward the rear of the courthouse. "That's where the Donner Party left from before...well, before that unpleasantness."

The story of the Donner-Reed Party had shocked the country. After leaving Springfield in the spring of 1846, traveling to California, the infamous party had been trapped by blizzards in the Sierra Nevada and some of them had resorted to cannibalism for their survival. He smiled ruefully. "That's about it for our visitor attractions."

Then he stood up straight and indicated the defense table with a wave. Lincoln and Logan were standing against the rail in discussion, while Quinn Harrison sat impassively, head bowed, hands clasped in front of him. "That and now Mr. Lincoln. For the last few months there have been people coming here to try him out." After a pause that Hitt interpreted as admiration for Lincoln, Kidd continued, "I read your words from the debates. All of us did." He tensed. There had been some complaint that the steno man had prettied up Lincoln's occasionally bumpy prose while ascribing to Douglas much coarser language. It wasn't true, of course, but the criticism had been hurtful to his professional pride. He steeled himself for that question, but he needn't have worried. "That was a Springfield man talking. He did us all proud. Should have gone to the Senate, too. But those Democrats..."

Lincoln was chuckling at something Logan had said to him. Seeing him in this setting, in a courtroom, he seemed quite different than he had on the speaker's platform. Hitt easily could recall the first time he'd seen him at the debates. There was something about him just standing on the platform in a group as preparations were concluding that had struck him then. Even while standing in the midst of that group, he somehow seemed not a part of it. There was an unusual sense of differentness about him. Nothing Hitt had seen in this courtroom had changed his mind about that. There was something unique about the man. That impression would be confirmed by this discussion with the court crier, Captain Kidd, who had been in the courtroom during the Almanac trial. "You know, he refused to take a fee from the family," Kidd told him. "I was standing right over there, right by that door—" he pointed with his index finger "—and I heard him tell the mother, 'Why, Hannah, I sha'not charge you a cent and anything else I can do for you, will do it willingly and without charge.' Then later, when some rascals tried to swindle her out of a small piece of land, he and Mr. Herndon did exactly what he promised."

Just then the door to the judge's chambers opened. A man Hitt did not recognize peeked out, and when he spotted Kidd beckoned him, indicating he should come immediately. Kidd sighed. "You'll excuse me. A pleasure meeting you, Mr. Hitt." With a wave of his now empty water pitcher he was gone.

As this brief break stretched into several minutes, Hitt finally stood; he began shaking his hands briskly from his wrists to increase the circulation of blood into his fingers, an exercise he had learned while working long hours as the official scribe of the Illinois State Senate. That done, he began moving about the courtroom to stretch his legs. This locomotion was something he used to enjoy, as it gave him an opportunity to feel the sense of the courtroom, but since he had gained his small celebrity for the work in the debates people knew his name and pointed him out, which made him uncomfortable. The trick, he had learned, was to keep moving. If he stopped even briefly, people wanted to discuss those debates or question him. As he moved around this courtroom, he overheard bits and pieces of conversations. Lincoln's name was bandied about, as was Palmer's. There were several mentions of Peter Cartwright in anticipation of the Reverend's appearance. "The confederacy..." was mentioned. In his experience, some courtrooms had felt angry, ready to explode; others had been light-hearted or even blasé; there was no universal feel to this courtroom, at least not that he could discern. If pushed, he would have described it as "intense," or "solemn." Even those bursts of laughter he had heard were subdued. There was serious business being done here, and everyone knew it.

He was just thinking with some pleasure that even with feelings at a pitch, there had been no loud arguments, no threats from supporters of one side made to the other. That was known to happen in courtrooms, especially when people had an interest to root for. Fights had been known to break out in the gallery. But that was not the feeling he got; in Judge Rice's courtroom there seemed to be respect for the process.

But as he passed the group of young men he had

seen congratulating Livergood, one of them, a slender man with an unkempt beard, stepped in front of him. In a manner clearly meant to impress his fellows, he said, "Hold up there, writer-man. You going to put down this story the right way?"

Hitt eyed him squarely. "I don't write stories," he explained. "I take down the words as they are said."

The rest of the group had turn to face him. One of them said, "All them sweet words of Logan and Lincoln, you mean." Around them other conversations had stopped as people turned to watch this encounter.

"And Mr. Palmer, too," Hitt said. "And Judge Rice and all the witnesses. My job is to record their words."

"Even when they're twisting up the truth?"

"It's not up to me to determine their veracity. I just put their words down." While Hitt did not feel threatened, it was an uncomfortable encounter.

"Vera city?" the bearded young man repeated, then literally scratched his head. "Is that like New York City?" He looked to his friends, who responded with supportive amusement, then continued, "I never heard of that one. Where's it at?"

"It's not a place," Hitt said. "It's a word that means the truth."

Another one of the group, a heavyset man with a pockmarked face, shouldered his way forward. "You really think what they're saying is the truth?" He shook his head with sadness. "Greek's dead and Peachy killed him. There wasn't no call for that. Greek was just setting things right for his family. He only meant to mess Peachy up, that would have been easy for him to do. Greek could handle things. Peachy should've fought him fair."

"No doubt 'bout that," the bearded man agreed.

The heavyset man sighed. "That's the truth of it. And there's nothing Lincoln, Logan, any of them can do here that's gonna make it right."

The fourth member of the group, clearly younger than the rest of them, finally spoke up. "Unless Greek has some Jesus in him and gonna rise up. But I never did see no angel in him."

Everyone laughed at the truth of that, and as they did, Springfield's sheriff, Joseph Pirkins, a white hat in his hand, his holstered six-shooter bouncing against his hip, pushed himself into the group. Pirkins was a tall man, almost as tall as Lincoln, and sturdy, with an artfully curled mustache adding to his intimidating presence. "You boys being kind to our visitor?" he asked pleasantly, but his purpose was unmistakable.

The attitude went out of the young men like air escaping a rubber balloon. "Yes, sir, Sheriff," the heavyset man said. "Just letting him know our thoughts about the affray."

Pirkins smiled and pointed a friendly finger at him. "Now, Charlie, what'd I tell you 'bout thinking. You got to cut that out. I promise you, no good's gonna come of it." The other young men turned their attention to Charlie, joining in the teasing. As they did, the sheriff took hold of Robert Hitt's elbow and escorted him away.

As the steno man started walking with Pirkins the slender man shouted after him, "You make sure you write it up good."

"They're not bad sorts," Pirkins said as they moved away, "but they can be a little boisterous." He pushed open the small gate separating the spectator gallery from the well and held it for Hitt to pass through. "We'll sure be pleased when this whole mess is over," he said, as much to himself as to Hitt. "It's been hard on all of us. The Craftons and the Harrisons, they helped build this city. Them and Abe." He followed Hitt through the passage, then let the gate loose. As it seesawed to a close, he took a chaw out of his sack and stuck it in his cheek. "You hear tell of the Long Nine?"

Hitt looked at him quizzically. "The Long Nine?" He shook his head. "No, I don't think so."

"Goes back more 'n twenty years now. Abe was in the state assembly. There was nine of them from Sangamon County, besides him one of them being Archie Herndon, Billy's dad, and it was the darndest thing. Every single one of them was tall as me. All nine of 'em."

"The Long Nine," Hitt said.

"Right. The Long Nine, that was them. Abe Lincoln may not have been the tallest among them, but he was the one they all looked up to. And when the legislature was fixing to move the state capital, it was these fellas who spoke up for Springfield and made it happen. Turned this whole city around. Made it an important place.

In addition to later becoming Lincoln's brother-in-law, the irascible Ninian W. Edwards was also a member of the Long Nine that managed to convince legislators to make the small town of Springfield the state capital. For many years his home was the social center of the city and it was there one Sunday afternoon that Lincoln met his future wife, Mary Todd.

"So around here people mostly like Lincoln, but a lot of them aren't too happy to see him in the middle of this one. There's gonna be some bad feelings, guaranteed, no matter how this turns out."

Suddenly the booming voice of Captain Kidd called the room to order. Sheriff Pirkins flashed a last smile. "You need anything, you let me know."

Hitt thanked him and returned to his seat. The courtroom was settling down. Lincoln and Logan were now engaged in conversation with Palmer and Broadwell. Clearly there was no animosity between these temporary adversaries. As Lincoln took his seat, he laid a comforting hand on Peachy Harrison's shoulder, then gently patted him twice, a gesture of reassurance.

John Crafton, Greek's brother and a participant in the fight, was scheduled to be the prosecution's next witness. Mr. Palmer would have to handle John Crafton with considerable caution. Crafton was more than an eyewitness, he had been an active participant in the fight. His brother had been killed. He, himself, had been cut badly and for a time feared for his own life. He had bled. A clever prosecutor could make jurors feel his pain. But while Crafton's testimony was essential to the prosecution's case, it carried with it danger for both sides. He brought substantial risk, both to himself and the case against Harrison, to the witness stand. His testimony was going to be quite dramatic, no question about that, and there was a fair chance he would draw sympathy from the jurors for his brother. Mr. Palmer likely would lead him to describe his brother lying on the floor, his life flowing out of him.

Crafton was going to testify to the facts as he witnessed them, as the brawl took place. It would not be favorable to Harrison. But Crafton had to be very careful not to overplay his hand. He had to tell the

THE LATE LLOYD OSTENDORF ARTWORK IS FROM THE LINCOLN COLLECTION OF THE LATE PHIL WAGNER, SPRINGFIELD, IL. IMAGE COURTESY OF WWW.ABELINCOLN.COM: ABRAHAM LINCOLN COLLECTIBLES DRAWINGS BY FAMOUS ARTIST

Lincoln offering a toast to six of his fellow Long Nine members on their success, as guests (reflected in the mirror) look on. State representative Ninian Edwards is sitting on the left while state senator Archer Herndon, the father of Lincoln's future partner and biographer William Herndon, is on the right.

truth as it was, without any embroidery or exaggeration, which meant also telling those parts that did not cast a positive light on his own actions that day. Lincoln and Logan would put strong meaning to those parts. In previous trials Hitt had transcribed, he had seen clever lawyers tie up witnesses in the claims of their own testimony, digging out the contradictions and soft truths, then using a man's own words to destroy his credibility. He'd seen decent people brought to tears in confusion, rendering their testimony worthless and laying an ax to their own case.

Mr. Palmer was going to have to find a way for Crafton to testify about what he had seen, and how he had suffered his own wounds, without making it obvious to the court that he and his brother were the aggressors, and that he had been slashed attempting to grab hold of Harrison, wrapping his arms around him, trying to restrain him while his brother apparently beat him. Walking that tightrope would require the skills of a funambulist like Jean-François Gravelet, who only two months earlier had thrilled America and Europe by crossing the thousand-foot-long Niagara Falls on a three-inch-thick rope.

And Palmer would have to do so without leaving loose ends for Lincoln and Logan to pluck. There was going to be some damage done to his case, he knew that, but it could not be helped. Rarely is a case without contradiction. But John Crafton would look directly at the jury and tell them that neither he nor Greek had brought a weapon; that Peachy was never in mortal danger; and that, rather than defending himself, Harrison was the villain. He had brought the weapon. He had killed Greek without sufficient cause.

Lincoln and Logan would have set very different goals for this witness. It had never ceased to amaze Hitt how two lawyers could make the same story told by the same witness come out so differently. The defense would use John Crafton to draw another picture entirely. They would use his testimony to lay the groundwork for the defense case: Quinn had been warned that Greek was out for revenge, that he had threatened to stomp his face, and so he carried a weapon to defend himself from the larger man. He had been attacked by the Crafton brothers; the well-meaning Ben Short was trying to pull him away. In those few seconds, fearing for his life, he pulled the knife with the white handle and struck blindly.

Judge Rice pounded his gavel three times, bringing the courtroom to order. Hitt dipped his pen as Captain Kidd called John Crafton to take the witness stand.

CHAPTER SEVEN

The first thing Hitt noticed was that John Crafton had been gussied up for his appearance. He was clean-shaven, and his hair had been greased into place. He was wearing leather pants and shined boots. But most glaring was the bright red handkerchief-sling in which he carried his right arm. In the highly unlikely event any of the jurors failed to notice it, as he took his seat and pulled the sling tight with his left hand he winced at the pain of the effort. Only when he placed his right hand on a Bible to take the oath did the steno man notice that Crafton's right wrist also was wrapped in cotton.

Unlike the friendly support that had accompanied Silas Livergood to the witness stand, as Crafton walked forward the courtroom was completely silent.

Palmer led him through the necessaries of his testimony. He lived at home with his parents, Wiley and Agnes Crafton, and his now-deceased brother, Greek, about a mile north of Pleasant Plains in Sangamon County. Yes, he knew Peachy Quinn Harrison. He last saw his brother after breakfast on the sixteenth of July. He had gone into Pleasant Plains while his brother had started to Berlinsville, for reasons he had not mentioned. He had next seen him at Mr. Turley's store, then soon afterward again at Short's drugstore.

Mr. Palmer asked Crafton to "Begin at that point at which you went to the drugstore and tell what you went there for, when your brother came and what took place." In Judge Rice's courtroom, in fact in many of the courtrooms of that time, questions like this that allowed for an often long and rambling answer were permitted. Giving a witness the freedom to have his or her say was a tactic respected by both sides and rarely was interrupted. There were moments when it served the opposition, as a too comfortable witness sometimes added unnecessary details that proved to weaken his own side.

Crafton's response was long and rich with those sort of details that must have sparked a storm in Lincoln's mind. "I went there as my business was for money," he began. "I called upon Mr. Short to know if any money had been left for me. He said not with him. He told me probably it was left with Mr. Hart, his partner."

It occurred to Hitt that John Crafton had been well prepared for his testimony. Everyone in Springfield would know Short and Hart were partners; mentioning that was unnecessary.

Crafton continued, "I asked where Mr. Hart was. He said in town somewhere. I said no more to him. Yes, I says, I'll wait for him here and I passed to the rear of the store and laid down on the counter." Mr. Crafton left the stand and, taking a pointer in his left hand, described for the jury precisely where he had lain. "Here was the counter beginning at the door on the west side and running all the way back to the south end except for three or four feet probably. I lay here at the south end of the west counter with my head to the south…" Crafton was a bit awkward with his left hand; it appeared he was unused to the sling, as his right arm got in the way of his pointing.

Hitt glanced at Lincoln, who once again had his hands clasped in front of his mouth, taking in Crafton's every word.

"When I went there nobody was in the drugstore but Short and Harrison. They were sitting right here on the west side of the drugstore near the front. When I went in and laid down I heard a noise at the door after I had lain there some minutes. I can't cite how long. I raised my head and saw three men in a scuffle. I saw my brother was one and Short and Harrison were the others. I made a spring and got to them.

I saw they were in a fight. They met me on the east side of the store about half way..." He slowly dragged the pointer along the diagram, and it was almost possible to see the brawlers moving toward the back of the store. "There I met them at the south end of the east counter. As I met them Mr. Short was pulling the boys backwards. He threw out his left hand and caught me. I told him to let him loose, that 'Greek could whip him.'

"He said, 'They shan't fight.' At that they all moved to the southwest on the west side." As he moved the pointer it slipped slightly; he almost grabbed it with his wounded arm but caught himself. "Here was a counter and by that counter stood some boxes, two or three, two probably. Mr. Harrison being on the east of Greek and Greek on the west of him and Short on the south of the two and me to the south of Short, and we all made a move to the southwest where these boxes were. There my brother fell on those boxes and in this leaning position Mr. Short pushed him backwards over that counter." He tapped the diagram several times to emphasize that position. "That's where I suppose my brother received his stab. I didn't see it. As Mr. Livergood interfered, Mr. Harrison sprang back a foot or two and made a motion to strike my brother again with the knife. Then I jumped over the counter (above it)... Mr. Short moved and I recovered. I jumped between Mr. Harrison and my brother. I threw my right arm to Harrison and he stuck me then with a knife and then jumped over here. I turned and in this position to catch him and he struck at me again. I saw he would use me up with the knife and I made to this east counter over here for the pound weights I knew were always there. As I started I met…" He mumbled a name that Hitt found entirely incomprehensible. As he had been taught, he left the space blank rather than guessing at it. He could almost hear the stern Mr. J. T. Ledbetter cautioning the class over and over. "Better to miss the stray word or name than pollute the transcript. Precision matters, my young friends, precision matters."

Whatever the name, "…he asked if I was hurt. I replied, 'My God! I am ruined!' I got to where the weights ought to have been, they were not there and I put my hands on the scales and threw them at Harrison. He was hollering, 'Jesus Christ! Have I got no friends here!' As I threw the scales I stumbled against a chair or stool. I picked it up and threw it at him and then turned around and threw two glasses at him. One of them I broke, the other I did not."

Palmer finally spoke up, asking, "Had you seen your brother from the time you left him until you saw him engaged in the fight?" His intent was clear: in his opening statement, Lincoln had claimed the Crafton brothers and Livergood had set out that morning to attack Harrison. Palmer was making it clear no such plan was afoot. This was a matter of circumstance.

Crafton made that point, replying, "I saw him at Turley's store, but I didn't speak to him." Then he repeated that, "I was in a hurry and didn't speak to him. I went out and left Mr. Harrison in the room after the fight. That is the last I saw of him. I had a pretty bad cut on the right arm in the muscle..." He made a show of attempting to raise his right arm but managed barely more than a few inches. "I have not got the use of it yet. The last I saw of my brother was on the boxes. As I turned I suppose he got up. I saw him next at Dr. Million's. He was then lying on the right side on the floor. I only just passed through the house then. I went to Dr. Million's office and laid down in the shade." John Crafton paused here and looked down at the planked floor for several seconds. When he looked up again, he did so with tears in his eyes. "I never saw my brother again, sir." In the gallery some minor stirring served mostly to call attention to the utter silence. A chair squeaked. A boot thumped on the floorboard.

After clearing the emotion out of his throat, Mr. Palmer continued, trying to disrupt Lincoln and Logan's attack before it might be mounted, "When you first saw Harrison and your brother and Short, in what position were they?"

"I think they were all right together. I think my brother was nearest the door. The first sight I got of him they had hold of each other. Mr. Short had my brother right by the right arm and Harrison by the left arm and Harrison and my brother were face-to-face. Short was to the right of my brother and to the left of Harrison."

"Did you take hold of either?" *Did you make it an unfair fight* was what he meant.

"No, sir. I got hold of Mr. Short by the left arm, by the left hand in his hand and my right arm about there…" he said, indicating his upper arm. By this time in most trials Hitt should have settled into what Mr. Ledbetter had referred to as his "work sense," in which he became little more than an empty vessel for the words to pass through, from the witness to the paper. In that state he wrote down his marks without conscious thought, *and* having as much meaning as a *stab*. But it was odd, in this case he was completely unable to reach that place; instead he listened hard as he worked, visualizing each description, intentionally ignoring the hems and haws and backtracks to get it right. To his surprise, he found himself caught up in the case. "…I caught him, him holding onto the boys with that arm."

"Did you do anything as you were coming back?"

"I don't think I did anything. I can't tell what I was doing. I was not pushing or pulling that I know of. We all made the move together. How it was done I can't tell. I saw the knife in Mr. Harrison's hand—saw it after he had cut him. I observed before he left the room that my brother was down and that he was hurt but didn't know that he was cut. He had his knee on one box and his arm on another box, one box leaning on the other, his knee on the lower and his elbow on the higher." Crafton again made a feeble effort to illustrate his point, but the sling prevented him from raising his arm.

John Palmer told Judge Rice he was done with his questions. The judge looked at the defense table and pointed: your turn. Slowly, as if burdened by thought, Abe Lincoln rose. Among his peers, Lincoln was considered quite skilled at cross-examination. He was known for never asking a question without a direct purpose, although often that purpose would not become clear till much later, and for having the rare ability among lawyers to stop a witness when he had what he wanted, rather than enjoying the sound of his own brilliance so much that he asked more than necessary. He greeted the witness, whom he appeared to know, perhaps having become acquainted while Greek was learning in his office. "Mr. Crafton, John," he began, "had you been at Short and Hart's store any but the one time that morning?"

"No, sir, only that one time. I had not been in before. I went there to inquire if some money had been left for me. I expected Bob Irwin, Robert Irwin, would leave it for me." Irwin? Hitt wondered, might that be the missing name? He took pride in leaving few spaces in his transcript.

"Short told you none was left for you?"

Crafton considered this, perhaps more than necessary for such a simple question. Hitt guessed he had been forewarned that Lincoln was skilled at laying traps for witnesses and was turning this one over. "It might have been Mr. Hart," he said, adjusting his earlier testimony. "I took my position on the counter and stopped for Mr. Hart. I saw nothing then until hearing the noise. I raised up and saw the three together."

Lincoln's voice had a heavy tone to it. "About that time did you hear either of the parties say anything?"

"No, sir, I think not. I don't think there was a word spoken. I spoke the first word, I think."

Lincoln looked puzzled, and asked, "You can't remember Harrison saying to your brother that he would not fight or didn't want to fight?"

"No, sir," Crafton said firmly, shaking his head, "I don't remember any such thing. I think I said the first thing spoken in the room. I told Mr. Short to let them loose, that Greek could whip him."

Having the benefit of John Crafton's earlier testimony at the Grand Jury proceedings allowed Lincoln to bring attention to small inconsistencies. "You did not add that Greek *should* whip him?"

"No, sir," Crafton replied, far more certain than he might have been, Hitt thought. "But I told him he *could* whip him." Hitt made a mark at the sign for "could," reminding him to underline it later, a good steno man's way of pointing out emphasis.

Having made his point, Lincoln moved along. "Did you see anything about the beginning or for a little while of Harrison holding onto one of the counters?"

"No, sir. He had passed almost to the end of the counter when I got to him, the south end of the east counter. It is only about halfway across the room. It don't go down and join… The opening is by reason of the eastern counter being there. I don't know what it is for. I saw them first at the door and when I got to them they were near the end of the counter. They were not at the west counter then. I did not see anybody holding on."

The timing and the movements, the west, the east, front and back, seemed to become a jumble. But Livergood had so clearly described Harrison holding onto the counter that Lincoln pressed the witness. "Were they so far from the counter as to leave no question about it in your mind?"

John Crafton admitted, "I suppose if a man had tried he could have reached the counter from where they were. I don't know whether that eastern counter extends down exactly half way, but it is about half way. I got up about that point of time and made the remark I told you of and got hold of Short's left hand and arm."

"And then you pulled on him, rather to pull him out of the fight?"

"I could not say whether I pulled him or not. I caught his arm and told him to let them loose. I held on until he pushed me backwards over this south counter. The whole bunch of us moved to the southwest. Quinn came east and he was pushing west."

Lincoln asked questions to confirm the movements in a professorial manner; inquisitive rather than challenging. "You moved along rather in a row and you came to the boxes before Greek?"

"No. I think Greek fell about the time I leaned

over this counter, pushed by Short. About that time I saw him fall."

"Tilting on the boxes?"

"Yes. The boxes were not very high."

Lincoln turned and glanced at Logan, as if trying to clarify something in his own mind. "The amount of falling then, was coming against the box with his legs? The balance was lost? He didn't fall flat?"

"No, sir, not flat."

A look of confusion crossed his face. "Then it was a sort of tilting?"

"I could not tell whether it was a tilting or pushing or falling. His knee, I think, was on one box and his elbow on another with his back against the counter that was higher than the boxes."

Hitt found it difficult to discern Lincoln's purpose in making this so important. It was said he picked his spots carefully, ignoring the fluff and bringing pressure to bear on the weakest points. But where was that point in this questioning? Still, Lincoln focused on it. "You were not interfered with by the boxes?"

"No, sir, I think not. These boxes were right in that corner and this counter running that way." His attempt to make it clear served only to further confuse the issue. "You see, I was south of all of them all the time and of course I would strike this counter and Greek would strike this one, the west one. I would get on this counter and he more westward. I was working more south than he was."

The line of Lincoln's questioning suddenly became

clear. John Crafton was claiming that his brother had been sliced as he fell onto the boxes, not while he was holding and punching Harrison. According to his testimony, Harrison was not defending himself, he was the attacker. Greek was going down when he was struck.

That statement had to be countered before it was allowed to sink in and do damage to the defense. Lincoln then asked casually, but in a firm tone that landed somewhere between a question and a statement, "So you did not see the knife when the blow was given?"

"I did not see the knife go into him."

Lincoln hesitated for a moment, looking down at his table as if to examine a document that was not there. And nodding. The silence was powerful, leaving John Crafton's admission to float slowly through the courtroom: he did not see the cutting. He was not an eyewitness. He did not know, really, whether Greek had been stabbed standing up holding Harrison or while falling over the boxes.

Finally, Lincoln resumed. "When you first saw the tussle in what way did Quinn and your brother have hold of each other?"

"Well, sir, I could not state exactly for the reason that all three were close together, but I think my brother has his left arm over Quinn's right one. His right one was held by Short."

"You have said Quinn and Greek were face-to-face?"

"They were when I got to them. As I said, they

were face-to-face from the time I got to them. They were face-to-face I think all the time."

Lincoln looked puzzled. After seeming to attempt to figure out this dance in his head, he finally came out from behind his table, hands clasped behind him, his gaze still down. Suddenly he loosened his hands and raised his right index finger, as if the thought had just occurred to him. "Did you see either of them strike at the other with his fist?"

Hitt covered a smile with his palm. A professional steno man never displays emotion. He certainly made no pretense to know the law, but even during the brief time he had worked in the courts he had learned the prime rule of cross-examination: never ask a question to which you did not already know the answer. Here was Lincoln demonstrating the proper use of it: there had been ample testimony on this point already; the only blow had been struck by Greek Crafton.

"I think I did," Crafton said, slightly raising his right hand. "I think with this hand. If you will give me a couple of men, I'll show you how it was."

Lincoln turned to face the prosecutor, saying, "I don't know whether it will do to risk myself, but I'll go in if Mr. Palmer will."

John Palmer accepted the challenge. "Am I to do the striking?"

"I'll make myself Greek," John Crafton said, missing entirely the sadness of that statement. But others in the courtroom caught it. Unable to use his wounded right arm in this demonstration, John Crafton reached up with his left arm and put that hand on Lincoln's shoulder, turning him to face the jury. He placed Palmer at his side, holding one arm. Then he stepped behind Lincoln, almost hidden from the jury's sight by Lincoln's bulk, and wrapped an arm around his body. "They were just this way and I saw Greek make a motion, but whether he hit him or not I could not say. That's the only lick I saw struck."

Lincoln freed himself and faced the witness. "I suppose when you told Short to let him alone, that Greek could whip him, that you had the ordinary meaning and wanted Greek to whip him?"

"I thought after they got into the fight that Greek could whip him."

"And you wanted him to?" There, there it was, Hitt knew. There was Lincoln throwing the shadow of doubt over Crafton's testimony. Reminding the jurors where his loyalty lay that day, and this day.

"Well… I certainly did."

That was Lincoln's final question. John Crafton walked out of the courtroom, the only sound the thump of his boots. As he went through the door, two other young men went with him, one of them throwing his arm around Crafton's shoulders. The prosecution called Daniel Harnett, still another young man of the town. All of them had grown up together and formed their allegiances. Harnett was smaller than the rest of them, but broad shouldered. A pair of black braces, crossing in front of his chest, held up new deerskin trousers. He took the oath and told the jury he lived at Pleasant Plains and he saw Greek Crafton and Quinn Harrison there on the morning of July 16. "I saw them after they were through with the fight. They were in the drugstore when I saw them. I had seen Greek Crafton before. I hadn't seen Quinn Harrison before."

Palmer let him tell his story. Neither Logan nor Lincoln saw a reason to object to any part of it. "Well, sir, I saw after the affray commenced. I went down to the drugstore from Mr. Turley's store and I saw Harrison strike once with a knife. The lick was towards John Crafton. That didn't hit him though, I think. That was the only lick I saw him strike. I saw John throw a pair of scales. I saw Greek. I didn't see him do anything." He hesitated here, seeing it again in his memory, but only now understanding the toll. When he started again, his voice was slower, more reflective. "He was lying on his side, I think right against a box. When I first saw Greek, Harrison was a few steps east of Greek, about as far from here to where Mr. Broadwell is sitting…"

As he had been taught to do, Hitt made a quick approximation of the distance, being quite careful not to be too specific: eight to ten feet he estimated.

"…he was facing north. Greek was west of him. John Crafton was facing Quinn."

"Did you observe any marks on Greek's person?"

"No, sir, not at that time. I can't say I observed anything peculiar in his appearance, but I didn't see him very plain. I was noticing John and Harrison. They were in the affray."

"How long did Greek remain?"

"A very short time, afore he retired out of the room."

That was it for Palmer's questions. This time Logan stood for the defense. "Mr. Harnett, did you hear anything said between the parties while the fight was going on?"

Harnett nodded as he responded, "I heard Quinn say something. I think he said, 'My God! Have I no friends here!' I think those were the words he used. I think that was after John had thrown the scales at him, likely he had thrown a stool and the scales too."

Logan cut down any remaining value his testimony might have had with his final question. "You didn't see a scuffle going on in which Greek was engaged?"

"No, sir," Harnett said.

"That's all," Logan told Judge Rice. Harnett was dismissed. Palmer's next witness was Frederick Henry, a tall and gangly young man who ambled pleasantly to the witness chair, a broad smile in place. Palmer led him to the crux of his testimony, the actions of the fourth of July, the morning of the picnic and almost two weeks before the deadly encounter. While listening to his story, it was difficult for the spectators not to think how well the day had started out, until it made its turn. "I first saw Quinn Harrison at his father's house. He and I were together going in a buggy to a picnic at Clary's Grove. We met Greek and John Crafton about a mile and a half or two miles from Pleasant Plains."

"What took place?"

"We stopped and talked with the boys a few moments and Quinn Harrison inquired where his brother Peter was, and Greek Crafton said he was down in the wagon getting some ladies to go to the picnic. Then Mr. Crafton said, 'I understand you have been giving Peter a lecture about the company he has been keeping heretofore?' Mr. Harrison did not know but he had said something about it. Mr. Crafton then said at some convenient time they would settle it. Harrison said, 'It didn't make a damned bit of difference to him whether he settled or not...'" For Hitt, the most challenging part of getting it down right was knowing when and where to put quotation marks. It often was difficult to determine if a witness was actually quoting someone or just putting their words into his or her own thoughts. Mr. Ledbetter had offered only little guidance on the issue, pointing out that "said" does not always mean the words were actually said. Mr. Ledbetter's best advice was that care should be taken not to put unspoken words in anyone's mouth. It was impossible from the context to know if these were Quinn Harrison's direct words. Hitt put down the quotation marks, but did not feel confident doing it.

"...Mr. Crafton then retired a few steps and pulled off his coat and said, 'We can settle it now.'"

A direct quote. Hitt wrote down his marks, feeling better. "Then Mr. Harrison said he didn't want to fight. I told them 'They couldn't fight there.' Harrison said 'He couldn't fight' or 'Didn't want to fight,' or something like that. Crafton then said if he ever cast any reflections on his character he would whip him. Harrison said, 'You d—d son of a b—h. I'll shoot you and pay for you.' Crafton then made an exertion, I believe, to get into the buggy and I pushed him out. He picked up a clod of dirt and threw it at Harrison and struck me in place of Harrison. He then made another attempt to get into the buggy. I pushed him out and started on with the buggy. Crafton said something again after I started on. Harrison then caught the lines and stopped the buggy to get out, and he says, 'D—n you, if you ever lay hands on me I'll shoot you,' or 'I'll kill you,' or something of that kind. That was all that occurred there."

Hitt was careful with his quote marks. This was a direct threat to life and woe would be the steno man who didn't put them down exactly as they were said, including the proper punctuation!

Palmer asked in his matter-of-fact way, "Do you recollect the precise language used by Harrison when he called Crafton a son of a b—h?"

This was another issue entirely. How to include bawdy language? Ledbetter said it was completely up to the predilections of each reporter, but preached the virtues of clean language. Personally, he had explained, he did not blush when such words were spoken in an open courtroom, but professionally he preferred to replace middle letters with dashes. Hitt had adopted that usage, and had encountered no objections to that decision. Fred Henry was given to such colorful language. "I am not certain," he replied, "whether he said to him 'You d—d spreckled-faced son of a b—h,' but he had something connected with it. I wouldn't be positive as to the words. Then I drove off and left Crafton standing."

Palmer set the stage for him, asking, "Did you see them together again that day?"

"Yes. They met once that day after that on the grounds at the picnic. Crafton came up and offered to make peace with Harrison. As near as I recollect, I could not tell his language exactly, but it was Crafton offered to make up the fuss, the dispute, and Harrison said he had nothing to make up. Crafton said he had cast some reflections on his parents by calling him a son of a b—h. Harrison said he did not mean to go further than himself. Nothing more took place."

Hitt glanced at Lincoln, who uncharacteristically was writing a note. His hand scrawled rapidly across a sheet of paper. Henry's seemingly easy testimony carried great danger to the defense. If the men had parted so well there was no case for Harrison to be afraid, no reason to arm himself.

Palmer asked his witness a final question. "At what time of day did you see John Crafton?"

"It was about eight or nine o'clock along in the morning. I would not be positive as to the time I saw the boys together at Clary's Grove, but it was the after-past of the day. It was about the middle or past the middle of the day. I think it was shortly before I started home. I have no distinct recollection as to the time of day."

As Palmer sat down, Lincoln stood up, ready to repair the damage that had been done. The afternoon heat had been growing in the courtroom, so as Lincoln got up he removed his jacket and placed it over the back of his chair. "Mr. Henry, did Quinn at the place where you first met Greek or at Clary's Grove in your hearing tell Greek that all he wanted of him was to let him alone, or anything of that import?"

Henry had the odd habit of holding his nose between his thumb and forefinger as he ruminated. "I could not make any positive statement," he finally decided. "It runs in my mind that there was something of the kind said. It runs in my mind that he said it either to me after we went away or he said it to Greek." As he continued, he gained more confidence in his own confusion. "I am not certain whether it was said to Greek or me. I could not make a definite statement. Since the question was asked I remember something of the kind being said. I could not remember whether it was said while he was standing there or while we were going home."

Lincoln made it easier for him. "Do you remember his saying at any time that he didn't want to fight?"

Henry remembered that. He nodded vigorously, glad to have a strong answer. "He said so I believe there when Greek wanted to fight him. I told him, 'We can't have any fighting here.'"

Lincoln nodded his approval. A good answer. He asked a final question, Hitt suspected, more to plant a thought in the jury's mind than needing the response. "Did you hear him say at any time that he, Quinn, was not able to fight Greek?"

"I don't remember that I did."

Lincoln told Judge Rice, "I've got no more questions" and turned to return to his seat. Henry grabbed his nose one last time as he looked at the judge, who dismissed him from the stand.

Hitt had never been in a courtroom that held the heat quite as well as this one. The windows of the courtroom were opened their full length to let the day inside and catch whatever breezes might sweep through, but the heat was just pouring in. He had taken to mopping his brow regularly. Judge Rice noticed him doing so and asked after his well-being. "I'm fine, your honor. Thank you," he replied.

Judge Rice asked the bailiff to draw the shades, hoping to cut down on the direct sunlight. And then he decided, "Tell you what people. Let's take a little break right here. Why don't you all go 'head and get yourself something to cool down, and we'll see you back here in twenty minutes." He punctuated that with a single sharp blow of his gavel, and left the stand.

CHAPTER EIGHT

After refreshing himself, Mr. Robert Roberts Hitt found a shaded spot beneath the Revolutionary oak outside the courthouse and set himself down to rest for a few moments. From his sack he took the dog-eared copy of Frederick Douglass's most recent book, *My Bondage and My Freedom*, and began reading, quickly finding himself lost in the horror of slavery that he described.

> This young woman was the daughter of Milly, an own aunt of mine. The poor girl, on arriving at our house, presented a pitiable appearance. She had left in haste, and without preparation;

> and probably without the knowledge of Mr. Plummer. She had traveled twelve miles, bare-footed, bare-necked and bare-headed. Her neck and shoulders were covered with scars, newly made, and not content with marring her neck and shoulders with the cowhide, the cowardly brute had dealt her a blow on the head with a hickory club, which cut a horrible gash, and left her face literally covered with blood. In this condition the poor young woman came down, to implore protection at the hands of my old master. I expected to see him boil over with rage at the revolting deed, and to hear him fill the air with curses upon the brutal Plummer, but I was disappointed. He sternly told her, in an angry tone, he "believed she deserved every bit of it," and if she did not go home instantly, he would himself take the remaining skin from her neck and back…

The sounds of spectators returning to the courtroom took him out of the book. He picked a twig off the ground and stuck it in the page for reference. He put the book back in the sack, brushed himself and joined the line returning to the courtroom. Douglass perplexed him. He had heard complaints that Douglass was given to exaggeration, but everything he had read seemed true to what he knew. What he found difficult to reckon is how someone like this escaped slave Douglass, raised up in that fashion, could master the written language. He, himself, had enjoyed the advantages of university and still at times found himself stymied by it.

The whole slavery issue confounded Hitt, as it did just about everyone with whom he discussed it. The morality of the issue was obvious, but getting out of it was so complicated. The economy of the South had grown dependent on it; its abrupt elimination would cause widespread devastation throughout that region. People wanted easy answers and there were none. Even those people who abhorred it could not agree on a solution. Lincoln's position had wavered; at one time he favored colonization, freeing all slaves to settle in Liberia or Central America, but later proposed accepting it where it existed until it slowly evolved out of existence while prohibiting its expansion into new states or territories. But still, this was a civilized nation and reasonable men had to be able to find a way out of this without killing each other.

As he took his seat in the courtroom once again Hitt looked at Lincoln, who was sitting alone at the defense table, somewhere in thought. Suddenly Lincoln looked up and caught Hitt's eyes and then smiled at him. Hitt acknowledged the greeting and quickly made to look like he was working on a paper.

The afternoon break had allowed the courtroom to cool a bit, but for the first time there were a few empty seats in the gallery. Mr. Palmer asked that John C. Bone be called. Bone was another young man from a well-established family, as the Bones claimed to be cousins of both James Knox Polk and Sam Houston.

He admitted seeing both Greek and Peachy at the July Fourth picnic. Then Palmer asked him if he'd spoken with Harrison about the problem. He had, he said. "I suppose it was about nine o'clock. He came to me on that day at a celebration over in Clary's Grove, near Tallula. He came and took me to one side…"

Suddenly, Hitt heard someone ask, "Who took you?" It took him a few seconds to realize this question had come from a juror. No one else in the courtroom appeared to find this unusual.

The witness looked at the jury box and responded, "Why, Quinn Harrison. I think his first remark was, 'Have you got your tools with you?' Says I, 'No, I never carry them to any such place.' He went on and said that he and Greek had had a difficulty that morning and he was afraid they would have still further difficulties that day. I made the remark to him, 'Quinn, we are all here and you and Greek shall have no difficulty here today. There are other times you can settle this difficulty,' or something like that.

"Then he showed me a knife that he had. He said the reason he wanted something of me was because it was a borrowed knife and the man wanted it. It was between four and one-half and five-inches long in the blade, a regular bowie knife, sharp in the edge, a small bowie knife. I think the handle was German-silver."

With ease, Palmer was putting the murder weapon in Harrison's hands. "Where are such carried?"

Clearly familiar with blades, Bone explained, "Different persons carry them in different places. They generally have a scabbard. This was hung in the breast in some way. I didn't see the scabbard. He put his hand in his bosom and drew the knife out and showed it to me."

Palmer handed his witness to Lincoln, who was sitting at his table. At some point, Hitt suddenly noticed, Lincoln had put on a pair of spectacles. He laid them on the table, then, without rising, asked knowingly, "In that interview between you and Quinn was anything further said?"

"Yes," Bone said, adding in one great rush of speech, "he stated he didn't want to have any difficulty with Crafton and didn't intend to. He might have said it a little further. When he spoke to me about the knife and about Crafton wanting to fight him, I asked him if he wasn't able to fight him, and he said, no, he was not and he had told him he would not fight him and didn't intend to and said if he jumped on him he would hurt him if he could."

Stephen Logan then spoke up, clearly attempting to make a strong point from John Bone's testimony. "He was afraid there would be further difficulty?"

"Yes, he was afraid there would be a further difficulty and this knife was a borrowed knife and the man wanted it."

Logan followed up. "He remarked that the man wanted it and he wouldn't have anything to defend himself with?"

"Yes. That's why he said he wanted something of me."

Logan made his point once again, so strongly that

Hitt wondered whether to put a question mark or a period on the end. “He wanted something to defend himself with, if he jumped on him?” A question, Hitt decided.

“Yes, sir.”

Logan said he was done and Bone made to get up, but Palmer stopped him and asked him to repeat precisely the language he used. Bone told him, “He said the reason he wanted the knife of me was that was a borrowed knife and the man wanted it.”

“Did he say anything else in that conversation?”

Bone shook his head. “No, sir!”

Hitt knew that Bone had been a helpful witness for the defense, as he offered shoring to their contention that Quinn knew of Greek’s threats and felt it prudent to prepare himself.

The prosecutor’s next witness was Edmund Crafton, Greek Crafton’s uncle, who said he knew both boys and acknowledged he had spoken with Quinn Harrison about the issue. Edmund Crafton was well respected in Pleasant Plains, running the most successful stable. He was known to have a good way with horses, and people trusted him with their care. “On the night of the Fourth,” he said, “setting in my room I was told by a citizen of Pleasant Plains…”

Palmer interrupted, “Never mind that. Tell me what *he* said?” His tone seemed unduly harsh, causing Hitt to wonder if perhaps the heat of the afternoon might be reaching him.

“Very well, Sir. Then I will tell you what he said.

On the morning of the fifth of July he came to my stable and after me and him had settled a little pasture bill between us, I said to him, 'Quinn, I understand last night that you have said yesterday going to the picnic, that you should have damned the name of Crafton and that you would shoot and kill them all and pay for them, and if you were not worth enough your father was.' And I said, 'I want to know whether you include me or my family or any?' Said he, 'I didn't include anyone except Greek Crafton.' That was the whole conversation except our own business." Mr. Palmer sat down and shuffled some papers. Stephen Logan spoke for the defense, asking the witness if anyone had been present during this conversation. Crafton told him that Dr. Million had been there. Logan wanted to know if Crafton had asked Quinn who he was referring to; he had, and he was talking about Greek. "That he should have told Greek Crafton that he damned the name of Crafton and told Greek that he would shoot or kill the name of Crafton. That was on the way to the picnic. I understood he should have said that. He acknowledged to the saying of that."

Logan spread his palms to indicate some confusion, then asked how Crafton believed it was meant.

"By saying he did not include anyone of the name except Greek."

"Is that your inference?"

Ed Crafton seemed to take umbrage at that suggestion. "No, sir, that's what he said. I just give you his words." He added emphatically, "I infer nothing."

Logan didn't seem to notice his attitude. "Dr. Million you say was present?" He was, and that was Logan's last question.

Hitt watched him as he returned to his table. He had begun to notice how little communication there was between Lincoln and Logan. Each of them seemed to have staked a claim to certain witnesses, and they clearly had confidence in each other as they neither offered nor requested advice. Occasionally one of them would lean over and whisper something to Quinn Harrison, who invariably nodded with understanding.

Twenty-six-year-old William Graham Purvines, Palmer's next witness, moved rapidly and purposefully to the witness stand, every firm step demonstrating he was a no-nonsense man. The Purvines family was well-known and respected in Sangamon County. Revolutionary War veteran John Purviance, as he then spelled it, had been among its early settlers, arriving there from Cabarras County, North Carolina, in 1820. The Purvineses were known to be hardworking people, and with kin had settled farms throughout the whole area. And like just about everybody else, most of them enjoyed good relations with both the Craftons and the Harrisons. In fact, William's sister-in-law, Frances, had married Peachy Harrison's older brother William after the death of her husband. Billy Purvines knew both of the fighters, and had spoken with Quinn Harrison in the afternoon of the picnic day. "It was between two and three o'clock, I suppose," he remembered. "Harrison said him and Greek

Crafton had had a little difficulty that day. He said he didn't want to have any trouble or fuss, I think, with him. He said though if he jumped on him, if he laid hands on him or crossed his path that he would cut his guts out; he would as soon kill him as kill a dog. That I think was as near the words as I can recollect. I don't recollect anything else said. That's all I got to say."

And with that he got off the chair and limped out of the room. As John Palmer called his next witness, Mr. Peter Livergood, Hitt intertwined his fingers and pressed down with the tips, bringing some relief. Hitt's right hand had begun to ache with fatigue, but there was nothing to be done about it. He did the daily exercises as he had been taught, the spreading and the bending, and when not assigned he did his practice letters, but there was no way to properly prepare for these long and busy days in a courtroom. Some of the veteran scribes preached the advantages of employing several pens of different weight and thickness, claiming even those slight variations seemed to make a difference as the day progressed.

Like the previous witnesses, Livergood was a resident of Pleasant Plains who knew all of the combatants. He told the prosecutor that he had gone to the Harrison house after being told of the fighting. Peachy's father, Peyton Harrison, "showed me where Quinn was and I went up to his room. I first asked him if he was hurt very bad. He said that he was. Then he asked me about the other boys, whether they were hurt pretty bad. I told him I thought they were hurt pretty bad, I didn't think Greek would live until sundown. I did not know how bad John was hurt. Quinn then made the remark that there was no danger, there was no danger of killing a hound. He said if his knife would have been longer he would have done his work up quicker and easier. I don't know which of these two words he used."

Palmer thanked his witness, then turned him over to the defense. Mr. Logan stood. As he did, it occurred suddenly to Hitt why he was sore. This trial was done more casually than those he had worked in Chicago. Rather than peppering witnesses with questions, jabbing for a single kernel of information or searching for slight inconsistencies, the lawyers simply allowed the witnesses to tell their story in their own words in their own way. Those trials he had worked in Chicago sometimes had seemed to him to be more a clash of lawyers proving their brightness than an attempt to get out the entirety of the event. There was an honesty about it being done this way that appealed to Hitt, even if it did make his hand hurt. He wondered briefly, as he dipped his pen, if this was a reflection of the difference in the way people led their lives in the big city and the country. He had no answer for that.

Logan wondered who else was there during Peter Livergood's conversation with Quinn. A working man named Whitehurst was there. Mrs. Harrison also was there, he thought. Unable to get that settled with any certainty, Logan reviewed the conversation, asking Livergood to confirm that Peachy Quinn Harrison had said there was no danger of killing a hound, presumably a reference to Greek Crafton, and if his knife had been longer he would have done it easier. "Yes, sir," Livergood agreed, "or quicker or easier." Minutes later, after more discussion about who was in the room, he added, "I asked if he was hurt. He said yes, he was hurt bad."

"Did he show any sign of it?"

"When I put my hand on his breast he said, 'Take it away.' He let on he was hurt. I expect I would have hurt him. He was lying there with his shirt bosom open."

Mr. Logan's voice was dragging, and he stopped to cough a bit before asking, "Did you see any marks on his side or breast?"

"No, sir. His hand was marked."

"Did you look to see any mark on his side or breast?"

"No, sir."

Mr. Logan told Judge Rice that was all his questions for this witness. Mr. Palmer then thanked all who had spent their day in the courtroom, and said he was resting his case. Judge Rice did the final business, announcing that the court would reconvene the following morning "about nine." Then he reminded the jurors, "Don't you go talking to anybody about anything you heard today, or what's going on in your mind about all this. We got a lot more to hear about this whole issue."

Bang! Dismissed.

The courtroom burst into life. People exhaled. Chairs squeaked. Conversations began. As voices were raised, the noise grew quickly into a near din. At the defense table Lincoln was carefully putting away his glasses and neatly packing a few pages into his silk hat. A man who previously had been identified to Hitt as Peyton Harrison, the father of the accused, came forward and leaned over Lincoln's right shoulder. Lincoln turned in his seat and seemed pleased to see him. Logan was talking to young Harrison, gesticulating broadly as he did. Palmer walked over to the defense table and said what obviously were some pleasant words to his adversaries. There clearly was no rancor between the parties.

Hitt closed his inkwells, cleaned his nibs and put them back in their case, blotted his latest pages and blew on the ink to make it fast, then leaned back and closed his eyes. It had been a long day, made even more difficult by the oppressive heat. After a moment, and taking several deep and satisfying breaths, he continued packing up his workspace. He had filled more than thirty pages with his markings, which would have to be transcribed. There were times he would immediately repair to an office and do that work while it was clear in his memory, but he had decided it was not necessary. He checked his timepiece; it was near six o'clock.

Coincidently, both he and Lincoln stood; Lincoln rolled down his sleeves and put on his jacket. After a few words with Logan and young Harrison, he left the courtroom with Peyton Harrison. He was stopped several times by spectators and exchanged a few words, the subjects far out of Hitt's hearing but it seemed pleasant enough. Once there was light laughter, which somehow did not feel inappropriate. Then he was gone.

While Hitt made his way to the Globe, Lincoln returned to his office, stopping along the way at Diller's Drug Store. And as was later recorded in his daily log, fulfilled the list given to him by Mrs. Lincoln. He purchased a bottle of the popular bedbug remedy, Dead Shot, a box of Wright's pills, a cathartic, pints of spirits and camphor and an ounce of glycerin. Mary Todd's list having been satisfied, he also purchased a small bottle of Guerlain's Imperial Perfume, a pleasant blending of bergamot, lemon, lime and mandarin, hoping it would lead to an untroubled night. But that was for later, when his work was done.

Herndon was there when he arrived at the office, doing the preparation for several upcoming trials. They briefly discussed the events of the day, both of them agreeing John Palmer had laid out his prosecution professionally if without much passion. He had demonstrated that Greek had died ugly, from wounds inflicted by Harrison, and that the Crafton brothers had come unarmed to Short's. He had done his job without leaving obvious openings for the defense. But that was what they had anticipated; they knew John Palmer, they knew the quality of his lawyering. But still, neither one of them saw any reason to change their own strategy.

ABRAHAM LINCOLN PRESIDENTIAL LIBRARY AND MUSEUM

Billy Herndon learned the law in the office of Logan and Lincoln, then formed a partnership with Lincoln in 1844 that would last until the presidential election of 1860. His three-volume biography of his former partner, Herndon's Lincoln: The True Story of a Great Life, *was published to great controversy in 1889, at least partially because he presented family stories critical of Mary Todd Lincoln, and even claimed Lincoln was suicidally depressed.*

Lincoln rolled up his sleeves again and began opening the firm's mail. It included several checks in addition to a clipping from the *Clinton Daily Public*, reporting on his appearance in the DeWitt Circuit Court the previous month. "The old familiar face of A. Lincoln is again amongst us," it read, "and we cannot help but noticing the particular friendly expression with which he greets everybody, and everybody greets him. He comes back to us after the electrifying debates, with all his blushing honors thick upon him; yet the poorest and plainest amongst our people fears not to approach, and never fails to receive a hearty welcome from him."

Both men got to the busy work of their business. Herndon wrote some checks. He drew up an application for client Josiah Day requesting a stay in a judgment against him for repayment of a $400.97 note. Lincoln drew up an agreement between Hutchinson's Cemetery plot owners reappointing John Hutchinson superintendent for another year. Finally Lincoln got to the business of stirring the political pot. He responded to an invitation from the Ohio Republican Party, agreeing to speak in Columbus and Cincinnati, but refusing other requests. He accepted an invitation to deliver the annual oration at the Wisconsin State Fair in Milwaukee in late September. In his work for the party he wrote to Richard Yates in Jacksonville, Illinois, telling him that he and Illinois senator Lyman Trumbull "anxiously desire" him to put himself up for office, telling him, "There is a strong desire for some—I rather think all—republicans here that you will allow them to run you for Congress in the Sixth District this fall." Yates previously had served two terms in the House as a Whig, but had been voted out of office by his proslavery district. Yates responded weeks later with gratitude for the offered support, but demurred, explaining he intended to run for governor, a strategy that would prove successful.

Lincoln then wrote a more complicated letter, this one to Ohio Republican congressman Thomas Corwin, whose support he would need if he was to make a serious run for the presidency. Corwin had written earlier asking Lincoln to explain clearly his position on slavery, wondering if it was true that he favored an abolitionist position for Republican candidates in Illinois. Lincoln responded, "Do you understand me as saying Illinois must have an extreme antislavery candidate? I do not so mean. We must have, though, a man who recognizes the slavery issue as being the living issue of the day; who does not hesitate to declare slavery a wrong, nor to deal with it as such; who believes in the power, and duty of Congress to prevent the spread of it." Before sealing it, he showed it to Herndon for his opinion. Herndon found it acceptably noncommittal, deftly stating the moral position rather than advocating a legal solution.

By the time the mail was done, the evening had become dark and cooler. A sweet breeze seemingly from nowhere had been pushing out the heat, making it almost pleasant. Herndon finally had closed the ledger, wished his partner good day, and taken off into the night.

Alone in his office Lincoln stretched out on the

floor, his back against the wall, and allowed his mind to wander through this case. As he once explained, "When I have a particular case in hand, I have that motive and feel an interest...in ferreting out the questions to the bottom, love to dig up the question by the roots and hold it up and dry before the fires of the mind." At times, he had found, if he did not try to focus his thoughts on one point or another, his mind took him to unexpected places and showed him a different path.

That had happened during the Armstrong trial, when for no obvious reason that even in retrospect he could figure out, he'd suddenly decided to consult the almanac and discovered the moonlight that night was insufficient for the eyewitness to have seen the crime. Making that discovery had been a moment of unrestrained pride; soon as he read it he knew his case was won and the boy was saved. But he expected no such moment in this case.

The conditions of this case were considerably murkier. The legal declaration of self-defense had been built on shifting grounds. There was no simple or consistent way of defining it.

He knew the accepted standards—he'd fought cases taking both sides of the issue—but he believed it was incumbent on him to delve deeply into the law so to prepare himself for whatever turns the case might take. He leafed through his tattered Blackstone Book 1, *Commentaries on the Laws of England: The Right of Person*, perhaps remembering the first time he had read through this book, by candlelight in a drafty cabin.

The historic right to claim self-defense was rooted in the great philosophers Plato and Socrates, Locke and Hobbes, who accepted it as a necessary right in a civilized, or organized, society. It appears that the ancient Athenians recognized self-defense in certain situations as "legal homicide." In 1651 the philosopher Thomas Hobbes had written in *Leviathan*, his explanation of the social contract, "If a man by the terrour of present death, be compelled to doe a fact against the Law, he is totally Excused; because no Law can oblige a man to abandon his own preservation."

In the early centuries the king had a sovereign duty to protect his subjects, the law required a common man to step back rather than meet the threat head-on. As the law in practice was sharpened, it required a man to retreat until his back, literally, was against the wall. The only exception to this was in his home, as it already had been recognized that a man's home is his castle. But when he was threatened, when he had nowhere to go, he was legally permitted to strike, and to strike with lethal force. With oil lamps by both shoulders, Lincoln reviewed his Blackstone on the subject: "The law requires that the person, who kills another in his own defence, should have retreated as far he conveniently or safely can, to avoid the violence of the assault, before he turns upon his assailant; and that, not fictiously, or in order to watch his opportunity, but from a real tenderness of shedding his brother's blood."

Lincoln sighed and rested the book in his lap. Blackstone clearly was not offering an easy application to this case. What conditions had to exist for a claim of self-protection to be made? Could young Quinn have avoided the fight? Could he have stepped back or even tried to run? Avoided it for how long? Forever? Should he have been compelled to live in fear? Allowed himself to be beaten? The thoughts began racing through Lincoln's mind. Harrison had been threatened. Where could he go to avoid this confrontation? Despite John Crafton's testimony, he was clearly attacked by the two Crafton brothers. It was essential that Lincoln establish that Quinn Harrison knew of Greek Crafton's threats, was afraid for his being and simply did what was needed to take himself out of danger.

The concept of self-defense wasn't quite as settled in parts of America, he knew. Americans were making the law to fit the conditions as they moved westward. It appeared that a lot of people were finding themselves in situations where there was no officer of the law to protect them or their family, so they took it into their own hands. As a result, the laws governing self-defense seemed a lot less restrictive than they once had been. In many frontier towns people often were satisfied to accept such a claim after a killing, knowing that someday they well might find themselves in an equally difficult situation. In Texas, for example, a new penal code passed only three years earlier practically favored the killer over his victim. Maybe people were stretching self-defense to mean self-protection, and the interpretations certainly were a little wobbly on the subject. Lincoln recognized the dangers in that, being publicly critical of "the increasing disregard for law which pervades the country; the growing disposition to substitute the wild and furious passions in lieu of the sober judgment of Courts." If this country were to persevere, he believed, it would have to defend the rule of law. Under the law in Illinois, a man had the right to defend himself when he was in jeopardy with no other clear options.

The problem, he well knew, was that under the existing law in Illinois a defendant could not take the stand on his own behalf.

So he could not tell the jury that he had firsthand knowledge of Greek's threats and was frightened of the larger, stronger man. He couldn't tell them he had gotten the knife to protect himself, never intending to use it. He couldn't explain that Greek and his brother had jumped him at Mr. Short's store and he was in fear for his very life when he struck out to push them away. He couldn't express his own sorrow that careless insults had come to this.

Lincoln would have to find a way to get that testimony into the trial through his other witnesses. That it would not be as effective he took for granted, but there was nothing to be done about that. The law was clear on the matter: according to the rules of evidence, a defendant was incompetent to testify in his own trial. It was believed that many people would say anything necessary, even under oath, to save themselves. A jury should not be swayed by subterfuge, sympathy or sorrow, so the accused could not have the opportunity to speak.

As happened sometimes, Lincoln dived deeper into the law, forgetting the passing time, forgetting Mary at home. He dug through the piles of law books on the floor until he found what he was looking for, a citation of the Grand Jury charge in the 1806 case, *Commonwealth v. Selfridge*. There, there it was. The first attempt in the new America to use self-defense as a justification in a homicide. An applicable precedent. Lincoln was reminded of the details with some delight: Mr. Benjamin Austin had purchased space in a Boston newspaper and published derogatory, if not defamatory, comments about attorney T. O. Selfridge. Selfridge had responded with publication of his own remarks, calling Austin "a coward, liar, and a scoundrel." Soon Selfridge learned that Austin had been making threats against him, promising to "attack me or flog me."

Lincoln knew of the case but had forgotten how directly helpful it could be here. While walking in the street the next day Selfridge had seen Austin's son, eighteen-year-old Harvard student Charles Austin, approaching him carrying a cane. Fearful that he was to be accosted with that weapon, Selfridge had shot and killed young Austin, then claimed to have done so in self-defense.

There was considerable debate about Selfridge's actions, with some people claiming Selfridge was never in jeopardy and that he had, in fact, shot and killed the innocent Austin as soon as he had seen him.

While there was some question about whether the doctrine of self-defense was proper in a manslaughter case, and how strongly Selfridge had actually made this claim, the Grand Jury had decided, "A man may repel force by force in defense of his person against anyone who manifestly intends, or endeavors by violence or surprise, feloniously to kill him. And he is not obliged to retreat, but may pursue his adversary until he has secured himself from all danger; and if he kill him in so doing, it is justifiable self-defense. But if the party killing had reasonable grounds for believing that the person slain had a felonious design against him, although it should afterward appear that there was no such design, it will not be murder, but will be either manslaughter or excusable homicide, according to the degree of caution and the probable grounds for such belief."

Selfridge had been found not guilty, although the reasoning of the jury during the actual trial was not written down or ever explained, but the law was built on such precedent.

Lincoln closed the book, keeping his thumb in it to hold the proper place. As he considered the way to use this information, he glanced around the office and was struck suddenly and unexpectedly by a memory. As he'd worked this evening, he had focused only on the winning of the case, not the flesh and blood of it. And now Greek Crafton had showed up in his mind. Lincoln and Billy Herndon had enjoyed having him work in the office. He had been a good young man; curious, punctual, industrious. Given his family connections, Lincoln had guessed he would have a productive career in the law ahead of him. Perhaps even politics; the boy certainly was smart enough.

Abe Lincoln was a realist; as Herndon described him, he "believed in predestination, foreordination, that all things were fixed, doomed one way or the other, from which there was no appeal." It was Mary who was intrigued with the possibility of spirit worlds. Mary had been known to consult seers. Lincoln humored her, although he did not share her curiosity. But he could recall at this moment the sound of Greek Crafton's voice and then his laughter, his booming laughter.

Greek had been boisterous; the exuberance of his youth had surged through him and was, perhaps a little too often for Herndon's pleasure, let loose in long and unrestrained laughter. Some people's laughter barely filled a barrel, but at times the walls of the large office had seemed barely able to restrain Greek's laughter.

Lincoln wondered to himself if he was not adequately mourning the loss. He stared at the roller chair, which had been left facing a bookcase, papers and books piled on the seat, and shook his head sadly.

He stood and straightened himself. He laid several law books on the long table and briefed through them. Within minutes he had found the other case he had been seeking out, an 1830 Tennessee case, *Grainger v. State*, which had set a new standard for self-defense. Grainger and a man named Broach had been drinking together at a tavern. Grainger left; no one had seen angry words between the men, but Broach pursued him, caught him and "struck him a violent blow on the breast." Grainger continued his efforts to flee, but Broach chased him. Finally Grainger stopped at a neighbor's house and withdrew the rifle he carried; he warned Broach not to come closer, threatening to shoot him—and when Broach continued to move forward Grainger did shoot and kill him.

He had been tried and convicted of murder and sentenced to hang, the jury believing his own life had not been at risk. Lincoln hesitated as he read those words, remembering his own case that had ended at the end of the hangman's rope. Grainger had been more fortunate; Tennessee Supreme Court Judge John Catron had overruled the conviction, writing that Grainger had "used all the means in his power to escape from an overriding bully," and shot only "to protect his person from threatened violence." Grainger, he added, had behaved like "a timid, cowardly man, was much alarmed, in imminent danger of a violent and instant assault and battery, and was cut off from the chance of probable assistance…"

In those words Lincoln recalled Harrison's plea, "Have I no friends here." He made note of this place in the book and put it with the others in his sack. It was time to go home.

CHAPTER NINE

John M. Palmer, too, had returned to his office. But before that, as was his practice while trying a case, he strolled twice around the perimeter of the square, running his hand along the railing that enclosed it, allowing the events of the day to settle in his mind. There were large bare spots in the green, he noticed, casualties of the dry summer. A game of town ball, the popular precursor to baseball, was in progress. He stopped briefly and watched the boys at play; like Lincoln, who actually played it, he thoroughly enjoyed it.

When he reached the office, the other members of his prosecution team, Jim White, John McClernand and Norman Broadwell, were there, coats off. White had brought fresh baked bread and green gooseberry jelly canned by his wife, and Broadwell had brewed a pot of tea, which was served as their evening meal as they sat at the long rough-timber table in Palmer's law office, reviewing the day and preparing their counter to Lincoln and Logan's defense.

The day had gone well, they agreed. They had made their case without leaving obvious paths for Lincoln and Logan to follow. They had proven that Harrison had carried a knife and killed an unarmed man while his own life was in no imminent danger, satisfying the legal requirements for both murder and manslaughter. But none of them deluded themselves into believing the case was nearly won. They knew well that this rope, their case, was only as strong as the weakest strand. And both Lincoln and Logan were expert at finding that thin thread. Then pulling it loose.

It was difficult to prepare for one of Lincoln's dramatic strokes. Once, the men at this table knew, he had been famously engaged in a case against his current co-counsel, Stephen Logan. After Logan had wowed the court with his argument, Lincoln had praised him effusively, but then pointed out that sometimes even the eloquent Judge Logan can be wrong, for "with all his caution and fastidiousness, he hasn't knowledge enough to put his shirt on right." Indeed, Steve Logan was wearing his pleated shirt inside out, a meaningless error that had allowed Lincoln to take much of the sting out of his presentation.

Palmer wondered aloud exactly how and where

Lincoln and Logan would make their stand the next day. The doctrine of self-defense had been pretty well established in law, he noted, and Harrison's back had been up against no wall. "I have it here," White said, reviewing the 1856 revised statutes. "'The use of a deadly weapon in self-defense is limited only to those events in which the danger is so urgent and pressing that in order to save his own life, or to prevent his receiving great bodily harm, the killing of the other was absolutely necessary.'" He closed the thick book. "They don't have it."

Broadwell smiled. "Well, you know Abe," he said. "He could sell you a mule, convince you it's a stallion and have you end up thanking him for the bargain!" The men laughed pleasantly—but knowingly, as all of them had experience with Lincoln's way of weaving words.

Palmer moved around the table refilling their teacups. "He can say whatever he wants," he offered, then from memory quoted John Adams, "but as Mr. Adams told us, 'Facts are stubborn things, and whatever may be our wishes, our inclinations or the dictums of our passions, they cannot alter the state of facts and evidence.' In this matter the facts do not favor Harrison."

"So where will Lincoln go?" White wondered. "Where's his opening?"

McClernand joked, "I doubt he's hiding another almanac up his sleeve."

"You, Jim," Palmer asked. "Where would you go?"

Broadwell interrupted. "Here," he said, tapping the cover of a book on the table in front of him. "Here. It's in your Adams, John," he explained, holding up Palmer's beautifully bound 1853 volume, *The Works of John Adams Esq, Second President of the United States*, for everyone's approval. "I think we all agree that self-defense shouldn't serve but—" he held up a cautionary index finger "—but there is an exception." He opened the book to a previously marked page. "I think we know the tragic events at King Street in March 1770."

McClernand was lost. "What the devil does the Boston Massacre have to do with this?"

Broadwell faced him. "It gives him his precedent, John. It's smaller than the eye of a needle, but he's fitted into tighter openings. We've all seen that."

American law had its foundation in the Crown's legal system. There certainly were exceptions, many of them done specifically to prevent the well-known excesses that had led to the bitterness, but even the courts recognized the similarities. Cases tried on the Continent wouldn't stand here, but there remained some confusion about reference to cases tried on American ground under British law.

"How does it play out?" Palmer asked, his curiosity now fully engaged.

"Five colonists were shot and killed by British soldiers that day," Broadwell reminded them. "Eight British soldiers were arrested to be tried for their murders. After every other attorney in the city turned them down, it was your sainted John Adams who agreed to defend them, John. No one knows why, he never explained it, really. But he took the job. The governing statutes were still British law, but self-defense was already well-trod ground. At trial Adams claimed that an aroused and armed mob had put the soldiers' lives at risk. 'A motley rabble of saucy boys, negroes and mulattoes, Irish teagues and outlandish jacktars,' is what he called them."

"Not too inflammatory, was he?" McClernand mused. "Rabble. Jacktars. My goodness."

Broadwell continued, "Adams convinced the jury that those soldiers had no choice. They had been abused by the locals for months. Snowballs with rocks in them were being thrown. They were being threatened with sticks. 'How was a soldier to respond,' Adams thundered. 'Do you expect he should behave like a stoic philosopher, lost in apathy?' The law said that it is the king's duty to protect his subjects, but old King George didn't happen to be in Boston that day. His great powers didn't stop the colonists from carrying nailed sticks, so he wasn't going to be able to do his duty. Adams contended those soldiers had a natural right to protect themselves. The jury agreed with him, letting six of them off."

Palmer had taken his seat. Broadwell had put on a fine performance, thundering like Adams himself might have done, but this argument seemed to support Harrison. After a moment of confused silence Palmer suggested, "I don't see how that serves us, Norman.

Perhaps I'm a bit dense, but this would seem to benefit our opponent. Isn't this the argument that Logan and Lincoln want to make?"

"Except," pointed out Broadwell, "except for Private Matthew Killroy. And right here is our case, John. It was Private Killroy who fired the shot that killed the ropemaker Samuel Gray. No one disputed that. But it was his claim of self-defense that was brought into question. Testimony was given that Killroy and Gray knew each other, and only a few days earlier the two men had engaged in a nasty dispute at Gray's Ropeworks…"

"Yes," Jim White said, grasping the concept.

"Hold up, there's more. Threats were made. A witness testified that Killroy had warned in anger that—" Broadwell paused and cleared his throat, then continued "'—that he would never miss an opportunity… to fire on the inhabitants.' Our Private Killroy was convicted of manslaughter. It was his own words that did him in." Norman Broadwell leaned back in his seat, quite satisfied his point had been made. With the backs of his fingers he nudged the Adams biography toward the middle of the table. "Thank you, Mr. Adams," he added.

"Harrison's own words," Palmer said.

"Manslaughter, at the least would be an appropriate charge," White agreed.

It was agreed that Palmer should hold this until it might be needed. Perhaps it would find a place in his summation. And if bargaining was to be done, then yes, they also agreed, manslaughter would satisfy. In fact, as Jim White suggested, given the passions in the city, it might be the best answer.

And please pass the jelly.

There were, of course, other matters of strategy to be discussed. There were two traps that had to be avoided.

The experienced Palmer had done an admirable job preventing any testimony showing that Quinn Harrison had firsthand knowledge of Crafton's threats. He had avoided putting on the stand any witness who might fill that gap. If Lincoln could not prove that Harrison knew personally of the threats, then he must be seen to have acted on the offense, not in his own defense.

The law seemed to leave little space for interpretation; a man had a duty to retreat until his back was against a wall. But the facts, Adams's stubborn facts, were clear: there was no wall at Harrison's back, real or imagined. Admittedly the question raised complicated legal issues, among them the value of hearsay. Should rumor have equal weight to a direct conversation? Are the whispered words of a friend sufficient to create a climate of fear? Was the letter of the law in conflict with the spirit of the law? Did the law protect Harrison or render him vulnerable?

As the men pondered these questions, a young man named Alexander Stevens, who was doing his law training in Palmer's office, arrived carrying iced steins of lager for the four men. "Direct from the icehouse," he said, "compliments of Mr. Farewell." It was a welcome diversion, and Stevens was lauded pleasantly for his initiative, with suggestions that a man who so well understood the needs of his elders was guaranteed to prosper. Palmer watched him moving about the table, so young and confident, so bursting with life, and he, too, was reminded of the tragedy that had them there in the office on a still-warm late September night. And as he sat there watching Stevens, he remembered that his first son, Benjamin Palmer, would have been about the same age as him. Would have been. It had been seven years since the consumption had taken him, but for the pain he still felt it might have been seven weeks. His nine children, who still lived their loud and happy lives, somehow had not been sufficient to fill that space. Ben still showed up in his mind at odd times, a memory that somehow helped him sort out his jumbled feelings. Ben had known both Greek and Quinn; they were part of the older boys who paid him little attention. If Ben had lived...but he hadn't lived. Like Greek.

So John Palmer knew the Craftons' anger, their pain. He would give them the best he could, the fair application of the law. It was not his job to decide the right or wrong of it. Rather his task was to be prepared, to make certain the law prevailed at all turns, whatever the outcome. He actually was pleased that Lincoln and Logan were for Harrison. That gave that young man his fair chance. Palmer had seen the outcome when incompetent lawyers had failed to provide their client with an adequate presentation of their case. He'd seen farms lost that should have been saved, and men imprisoned whose guilt remained questionable. He'd seen a man hanged protesting his innocence with his last breath. It wasn't always a fair system; it certainly wasn't always a good system. Even in this enlightened time too often the verdict depended on the cleverness of the lawyers to twist the law. And it was well-known that there were many lawyers who were not above doing so to gain their own advantage. But in this case there would be no trickery. In this case the lawyers were seeing to it that the law was done right. Quinn Harrison was getting his fair trial as he was guaranteed. The Crafton family was receiving justice, whether it gave them solace or not.

Palmer placed his emptied glass on the table, dismissing Stevens before he might refill it. There was a final issue to be discussed. It was no secret that the strongest witness for the defense was to be the Reverend Peter Cartwright. His appearance would make the day a difficult one for the prosecution. His testimony had to be blunted without his character or integrity being challenged. "So what's to be done about Uncle Peter?" he asked.

White responded lightly, "When he asks Judge Rice to join hands with him and pray for his nephew's deliverance, I say we object vehemently!"

The other lawyers smiled weakly in recognition of the problem. Then McClernand focused on the obvious. "He's the boy's grandfather for Go—" he caught himself in time to prevent even the touch of blasphemy, making the necessary correction "—for *goodness'* sake. The jurors know the connection..."

"But they also know Peter would never lie," Palmer said. "Not after swearing an oath. Not even to protect himself from the devil. Whatever he says, they will believe him." He admitted, "I'll believe him."

Broadwell added, "Then we have to prevent him from saying anything of consequence." He leaned forward over the table. "Let him be old pious Peter, the Lord's Plowman. Let him pontificate if he so chooses. Let him tell us all of his grandson's good deeds. The jurors will expect nothing less from him.

"But it isn't the quantity of his words that can do us damage. It's obviously the story he told at the hearing, Greek's dying words of forgiveness." Here he pitched his voice and did a whiny imitation of Cartwright's testimony at the inquest, "'I forgive Peachy and you should, too. It was all my own doing.' That's where real damage might be done."

Jim White made the telling point. "He isn't at a pulpit, though, is he? He'll be sitting in the witness chair. He's in our church, in our pew and the rules are different." He considered the legal options for a moment, then said flatly, "We can stop him. This is about the law, not the Lord."

Palmer said it first: "*Mortui non morden.* Dead men tell no tales."

Broadwell countered, "*Nemo moriturus proesum-*

itur mentiri. A man will not meet his maker with a lie in his mouth."

The statutes governing the admission of a dying declaration were loose around the edges. While this was not something generally studied during a legal apprenticeship, the question did arise surprisingly often in criminal trials, and everyone at the table was familiar with Blackstone's position on the matter. Mostly, secondhand testimony, hearsay, is not permitted to be heard in a courtroom. But one significant exception to that rule is the statement of a dying man. The law might be traced back to the twelfth century, where it was first recognized that no man on his deathbed would risk being condemned to the eternal fires of hell for telling a lie with the last breath. Through tradition it had come down through the centuries that dying declarations were only to be admitted in cases of homicide, where the circumstances of the death are the subject of the declaration. Blackstone stated plainly, "The general principle on which evidence of this kind is admitted, is that it is in declarations made in extremity, when the party is at the point of death, and where every hope of this world is gone, when every motive to falsehood is silenced…

"The statement of the deceased must be such as would be admissible if he were alive and could be examined as a witness, consequently a declaration upon matters of opinion, as distinguished from matters of fact, will not be receivable…"

Going further, Blackstone stated, "Dying declarations in favor of the party charged with the death are admissible as evidence, equally as where they operate against him… The question whether a dying declaration is admissible as evidence is exclusively for the consideration of the court."

It was White who finally said what they all believed to be true: "Well, if it's up to Judge Rice, this shan't be a problem for us then, should it." Judge Rice and Lincoln were well-known to have had their differences. While Judge Rice would give Lincoln the benefits of the law, he also would do him no favors.

"But there is a precedent that they'll find useful," Broadwell said. "The massacre again. Among the men shot that day was the Irish immigrant Patrick Carr. On the day before he breathed his last he told his surgeon, Dr. Jeffries, that the soldiers had been provoked by the mob to fire in self-defense and that he bore no malice against the man who killed him. They said the prosecutor, Samuel Quincy, was mad as a March hare that the judge allowed Jeffries to be heard, but could not sway him. Though the two judges reminded the jury in their instructions that the dying man had sworn no oath when he made his statement—" and here Broadwell read from the same paper he had consulted previously "'—But you will determine whether a man just stepping into eternity is not to be believed, especially in favor of a set of men by whom he had lost his life.'" He laid down the paper.

Palmer said with intended levity, "Of course those judges didn't know Mr. Abraham Lincoln!" And when the amusement had faded he continued, "I think there's some good in there for us, too." Should Judge Rice allow Greek's dying declaration to be heard as evidence, they would counter with equally strong rebuttal witnesses. Dr. Million, who had cared for Crafton and therefore had spent considerable time at his bedside, would testify that he never heard his patient make anything like those declarations, and the esteemed fifty-six-year-old founder and proprietor of the village of Pleasant Plains, Jacob Epler, who would swear that rather than exonerating Harrison, he had heard Greek angrily vowing revenge until the time of his death.

So their strategy was set; they would continue to contend that Quinn Harrison had no reason to arm himself with a deadly weapon, they would fight tooth and nail to prevent Cartwright from describing Greek Crafton's dying declaration, but if it was admitted they would counter it with Dr. Million and Jacob Epler. It had been a long but productive day for the four of them, and they were pleased to finally put down their work. They wished each other well and walked into the now balmy night. Palmer was last. He straightened up the table a bit out of habit, unnecessarily as there was a woman for that task, put out the candles and closed the door; then he straightened his shoulders and walked slowly toward his noisy house with that one empty room.

By this time in the evening, Robert Hitt was finishing a glorious supper. The Globe's cook had prepared spicy lamb chops, a fitting punctuation to Hitt's long but satisfying day. He'd left the courtroom intending to soak his throbbing hand then transcribe his notes; it would be a wearisome task but it had to be done. As he paused on the courthouse steps for ten deep cleansing breaths, a trick he'd learned that provided immediate renewal, he had been approached by reporter Solomon Wolfe of the *State Democrat.* The reporter Wolfe said his readers would be delighted to hear his opinion on the progress of the trial, as well as his reaction to the growing city. Hitt had tried to beg off, pointing out he was simply a quick transcriber, a man whose personal opinion wasn't worth the newsprint.

Wolfe had disputed that, explaining that R. R. Hitt, the famous law reporter, was known to readers from his work in Chicago trials as well as the Senate debates, and his presence in itself cast added importance on the trial.

Hitt had refused the bait, citing a nonexistent rule of his profession that steno men must never have opinions on the words they are transcribing, as admitting any bias would throw into doubt the accuracy of their work. As for the city of Springfield, on that subject he was willing to talk. It was a lovely place to be, he said, full of friendly folks who welcomed a stranger, and a place where all the modern conveniences could be found. He certainly hoped to return soon, without the press of his work, so he might enjoy all it offered. With that, he wished Wolfe a good evening and continued on his journey. But as he walked in the dusk (the blind man's light, as the British referred to it), his cheeks flushed, and he had to admit he had enjoyed even that small bit of attention.

That encounter was still in his mind when he reached the bed and board. Before the bell above the door had ceased its jingling, the stout Mrs. Sarah Beck, the Globe owner who supposedly ran it with her thus-far-never-seen husband, informed him that Mr. Thomas had invited him to join his table this evening. Soaking would wait, he decided. Actually, some additional rest before doing his work would be a good thing for his hand. Hitt refreshed quickly. Several tables had been set up in the rear garden, a place in which Mrs. Beck clearly and for obvious reasons took pride. While some of the annuals had lost their blooms, there remained sufficient color and variety to please the eyes.

The night was unusually light for this hour, and the colors of the sky bathed the entire room in the most unusual night shadows he had ever seen. It was if he were looking at it through a cloud of intensely bright strokes of red and orange, brushed against a powdered-blue background. As he walked into that cloud, he felt the pleasure of anticipation coursing through his body. Several townspeople had come to the Globe for the evening meal, and their lively conversation gave the place a slightly festive air. James Thomas was already seated at his table and stood to greet him. Thomas extended his hand to him, saying, "I'm so pleased you could join me. I suppose it was quite a day." Indicating the room with a sweep of his hand, he added, "Seems like it's the only thing anybody wants to talk about."

As Hitt sat, Mrs. Beck appeared and poured a glass of wine for him from the open bottle on the table. "It was—" he hesitated "—a very interesting day," he finished.

"Would I be imposing if I asked about it?"

Hitt smiled at that Southern politeness. It was a cliché, of course, but in his encounters he had found it to be true. Southerners did have a different air about them. They were not at all like people he knew in Chicago, who would dive straight in to a subject. But there was a second part to that cliché: don't mistake that charm for weakness. "No, not at all," he replied.

Actually, the invitation from James Thomas to join him for dinner was not at all unusual. Most people who traveled alone enjoyed sharing good conversation and a fine meal with people they met on the road. That was why most inns had long communal tables. Hitt began describing the highlights of the day, glad to be free to do so. As long as he voiced no opinions or revealed nothing private Mr. Ledbetter would have no objection. Initially he hesitated to repeat Dr. Million's graphic description of Crafton's wounds, it wasn't proper dinner talk, but as the wine loosened his resolve he provided most of the details. Thomas grimaced when he repeated the doctor's testimony about pushing Crafton's intestines back inside his body. When describing Lincoln playing Greek in the recreation of the crime and John Crafton's admissions and claims, Hitt found himself becoming particularly animated at which point he paused and reverted to the more dispassionate tone he had employed at the outset.

Hitt went on until the Italian cheese was served, and then the conversation drifted. He politely directed several questions to Thomas, learning he worked as an accountant for Wells Fargo, and he was in Illinois to explore the financial consequences of a proposed merger with the Butterfield Overland Company. Thomas did rattle on for several minutes about the exciting prospects of a national mail service and how that might help bind the nation together. Hitt feigned interest, but began paying more attention as Thomas began gradually easing the conversation toward Mr. Lincoln and then, surprisingly, to the great issue of the time.

Gradually, as Hitt savored his beefsteak and fried onions, Thomas began revealing more about himself. He did so slowly and, Hitt realized, with intention. These facts didn't leak; he wanted him to know them. By the time he was finishing his blackberry pie and coffee, he had put together a fuller picture. James Thomas was a son of Charleston, South Carolina. He had been traveling extensively for Wells Fargo for quite some time, trying to expand that company's reach by merger with other transporters, and been as far north as Boston. His parents had remained in Charleston, his father being somehow involved with "the land business," and while he had grown up with servants in the big house, he was perhaps intentionally unclear about whether they were paid help or slaves. Hitt, of course, could not ask. Thomas had an older brother who had graduated from West Point and was presently serving in the west. He told an amusing story about his brother, an officer in the Eighth Infantry, having recently participated in an odd trial to determine if camels were the equal of mules in transporting military equipment through harsh territory. Both men laughed lightly at the concept. Hitt told his own story. He had been born in Urbana, in Champaign County, Ohio, but his family moved to Mt. Morris, in Illinois, when his father, the Reverend Thomas Smith Hitt, was hired to minister a small congregation. Both men were pleased to discover they were about the same age. They talked around the slave issue until it became clear that neither of them held radical views. And then their mutual distress became clear.

They talked quietly as the night got darker, although still retaining an eerie glow, and the garden emptied. Around them serving people were clearing the tables, occasionally clacking plates together. For months now the northern newspapers had been filled with stories about the intractable South, determined to defend slavery even if it meant destroying the nation, so Hitt was pleased to hear firsthand a Southerner's report. It appeared, according to Thomas, that few of them wanted this fight, but were as perplexed as Northerners about how to stop it. He had seen and heard the zealots, he said, from Boston to Atlanta.

Both sides seemed equally convinced of the righteousness of their position.

Then Thomas began wondering aloud about Lincoln. While he was still very much the dark horse for the presidency, he had begun to attract interest in the South. The hard positions of Douglas, Seward and Chase were well established, but less was known about Lincoln beyond the fact that he was described as a man of the people. It was accepted that he favored ending slavery, but clearly he hadn't settled on a mechanism. In fact, Thomas said, "He sounded like a politician, a thus far lot of distant thunder but no sign of the storm." Was he a candidate the South could accept if not embrace? What was the extent of his commitment? How far would a "country lawyer" be willing to go to keep the union? If it came to it, would he fight? In so many ways he was an enigma.

"He claims to be against slavery," James Thomas continued, "yet he defended the legal trappings of that institution. We know where Douglas stands, he certainly tells us often enough. And Seward and Chase too have been outspoken. But your Mr. Lincoln..." He let his voice trail off. "Who is he?"

Hitt shook his head. "He is not *my* anything," he protested, strongly defending his professional neutrality. "I write down his words, whether I subscribe to them or not makes no difference."

"Springfield is Lincoln," Thomas retorted. "Lincoln is Springfield. He was shaped here. I would never deem to ask if you support him, Mr. Hitt, that is not my business and I beg your pardon if you thought that. But you were there for those debates. You've seen him at work in the courtroom. I am curious how you see the man."

Hitt sipped the last of his coffee. He considered the question, how did he see the man? The only person to whom he had stated his opinion was Thomas Jefferson Brown, his closest friend from his growing up days in Mt. Morris. They had been discussing the debates with Douglas, and Jeff had asked for a description. He wanted to know what it felt like to be there. Against his better judgment, Hitt had revealed his personal feelings. "I am inspired by him," he admitted.

He then told him a story. The third debate had been held in a town in southern Illinois, called Jonesboro. He traveled there with Lincoln. This was a very strong proslavery region known as "Egypt," and there were rumors that men were coming there from Kentucky and Missouri intending to stop Lincoln from speaking. When some people advised him not to speak, Hitt remembered Lincoln responding, "If only they will give me a fair chance to say a few words, I will fix them right."

A stand had been constructed in a grove near the edge of town. Lincoln and Hitt had spent the previous night with several others, sitting in front of the Union House in Jonesboro Square, watching the passage of Donati's Comet. The following afternoon, Lincoln began by confessing he had been warned that people there might make some trouble. "I don't understand why they should. I am a plain, common man, like the rest of you... Don't do any such foolish thing, fellow-citizens. Let us be friends, and treat each other as friends. I am one of the humblest and most peaceful men in the world, would wrong no-man, would interfere with no-man's rights. And all I ask is, having something to say, you will give me a decent hearing." And so charmed by his honesty, Hitt explained, they listened with respect. It was an especially significant meeting, Hitt continued. Before then Douglas had accused Lincoln of saying one thing in northern Illinois and something quite different in the southern part of the state. After that day, such claims became much harder to justify.

At first the debates had been more about Illinois's incumbent senator, Stephen Douglas, who was already a national figure of considerable importance. But as they proceeded, Lincoln, aided in part by Hitt's accurate transcription of his words that had been published and discussed throughout the north, had gained in recognition and stature until the men stood on the stage as equals. It was Lincoln himself who had requested the assistance of Hitt. The men had met in a Chicago courtroom as Lincoln defended a railroad from land claims he considered unjust. Hitt had stood out as a man in a profession that seemed to favor women. The curious Lincoln had asked him for an explanation of his work, and after a brief demonstration of how words were turned into symbols that would later enable those spoken words to be committed to paper, he had made a point of acknowledging Hitt each time their work brought them together. When the debates were scheduled, Lincoln had invited him to work with him.

It provided Hitt with a vantage point to history. He always had a reserved seat as Lincoln seemed to grow taller and bolder with each debate. He had come to appreciate the measured words of both men, both of whom were keenly aware of the impact they might have on the coming clash of regions. The debates centered more on the extension of slavery into new states and territories rather than its complete abolition. Hitt didn't feel either man had won, in a traditional sense, he'd told Brown. Each man had faithfully represented a position on the matter, but there seemed to be so many positions that there was no absolute right or wrong. Both Lincoln and Douglas had gained the support of those people who already agreed with their positions. At the end of the seven debates the situation was no closer to a resolution and, if anything, positions were hardening.

He centered his empty cup in the saucer. "Abe Lincoln is a complex man," he finally said. "The lawyers talk about his respect for the law and his strong sense of justice. They enjoy teasing him about it. Do you know he turns away clients if he doesn't believe in their case?"

"No, I did not. As I've said, there is too little known about him in the South. We know he's against slavery, what we don't know are his intentions."

That was the significant question. The South was openly discussing secession from the union. The word *war* was being tossed around as if it had no cost. "No one can answer that save Mr. Lincoln, and he is not about to do so," Hitt responded. "But there's no jelly in him, I can tell you that." During the debates, he had heard many stories about Abraham Lincoln. Among them was the fact that he preferred to walk away from a scrap, but when pushed hard enough he would make a stand. And when he did, he won. He always won. He told James Thomas several stories he had heard about Lincoln dispatching bullies. Although in appearance he was thin rather than muscular, the tales of his strength bordered on legendary. He knocked down three bullies in a fair fight. He carried six hundred pounds of logs. By himself he pulled a carriage out of the muck. "He did what was necessary, was the way it was told to me. And once started, he finished. I believed that was a fair description of him." Hitt quickly realized that he had offered far more overt support for Lincoln than he had intended to do.

"Now let me ask you, Mr. Thomas, is it this trial that has brought you to Springfield? Did you come to see Lincoln?"

"No, sir," he said, still pleasantly. "Not at all. I am here on Wells Fargo business. It is truly a fortuitous coincidence. Like so many of my brethren, I am simply curious about the depth of his belief, which like most politicians he has done an admirable job disguising. This election is going to affect all of us in ways we cannot yet determine. As I'm sure you know, a trial is a great show, but if he were to become a serious candidate, we would be more interested in seeing what he's like backstage."

Hitt found it necessary to correct him. "I believe you are doing him a disservice," he said. "As he once pointed out, if there was another man under his skin would he choose to wear that face?" They laughed at that. "No, say what you will about him, perhaps there are lawyers who have no problem bending their integrity or belief when necessary, but his actions during a trial are an accurate reflection of his character."

Thomas then offered his own reading of Hitt's remarks. "So then it makes no difference if his side should win or lose?"

Hitt seemed surprised at that, so surprised he responded with an unusually bold statement. "Oh of course it does," he said. "Winners win."

CHAPTER TEN

The oppressive heat had put Judge Edward Y. Rice in a sour mood, as did the knowledge that he was going to have to spend much of the day arguing the law with Abe Lincoln, Steve Logan and John Palmer. Lincoln may not have been able to cite the statutes by number, but he knew what the law said, and he was never too shy about reminding judges of it. Logan often reveled in a display of his own brilliance. And Palmer was a stickler for citations, precedents and the smallest details of legal history.

Judge Rice had stripped to his undergarments before buttoning the robe, trying to make himself as comfortable as possible. No one would know that though, not even his clerk. Judges had been wearing the black or dark violet robes since the passage of the Judges Rules of 1635 and no one ever wondered what was worn under it. *Well, maybe they do in Scottish courts*, he mused, giving rise to his first and isolated smile of the day. But here in the city of Springfield, the state of Illinois, and in the United States of America, no one paid much attention to what you wore or, as in this case, didn't wear. Far as Judge Rice was concerned, there were three advantages to the judge's robe: it kept you warm in the winter, cool in the summer and putting it on was like coating yourself in respect. It didn't seem to matter what man was wearing it; it was that robe that brought with it respect.

While wearing a robe was not an absolute requirement, Judge Rice chose to wear one when sitting in Springfield, as it lent decorum to the courtroom. He believed it was the majesty of the robe—combined with a good snap of his gavel—that shut up quarrelsome lawyers when he needed them shut up. Of course out on the circuit he dressed less formally in view of the casual surroundings, as well as the reality that packing and carrying the heavy robe from town to hamlet added an unnecessary burden. And keeping it clean in those places where mud floors were more common than wood was almost an impossibility. So most judges wore whatever clothes they were carrying with them, depending on the aura of justice to cloak them with respect.

Even after all his years on the bench, the irony of

courtroom life continued to confound Judge Rice. People wanted to believe that donning the robes of office and sitting behind a high desk guaranteed some kind of uniform application of the law to every courtroom, which was no more true than a sailor claiming that every storm was the same. Even he had to admit that he brought a different temperament into the courtroom every session, and as much as he tried to avoid it, there could be no doubt that those feelings were reflected in the way he ran the proceedings. Woe to the attorney who objected too vehemently the morning after Mrs. Rice had kept him up half the night suffering from the vapors. On the other hand there were those lucky men who were there when he came to court with a glint in his eye. Not that it happened too often anymore.

He stood before the mirror and straightened his robe. He lightly powdered the top of his head to prevent sweat from running down into his eyes and coughed his throat clear. It was time for work. And as he settled gingerly into his cushioned chair behind the bench and looked out at the overflowing courtroom, he knew for certain this was not going to be a good day to mess with him.

For Robert Hitt it already had been a busy morning. After his pleasant dinner with Mr. Thomas, he had returned to his room and worked deep into the night. After only a few hours' rest he had been at the telegraph office when it opened at 6:30 a.m. to send his transcription of yesterday's proceedings to Chicago. The telegraph operator informed him that some strange disturbance had interrupted traffic and that he could not say for sure when the long message would be transmitted. He literally scratched his head as he complained that it was the craziest thing he had ever seen; when the telegraph failed, he had turned off the power—and yet his key had continued to operate. Hitt smiled at the improbable story, and asked the man to continue his efforts on his behalf.

That unexpected interlude had given him time to enjoy the morning newspapers, and to his surprise, and pleasure, he had found himself mentioned in a broader story about the trial by the reporter with whom he had spoken.

A second story was more sensational, offering some of those "gruesome" details and speculating on which side had best made its case. But Hitt paused a minute to enjoy the mention of his name, as well as hoping that none of his colleagues, who most certainly would frown upon the attention, saw it. He read quickly through the other newspapers, which essentially reported the same events. It was easy to determine which paper favored which outcome by the size and the names in its headlines.

He noticed a brief story on the front page of the *Journal* that explained the difficulty with the telegraph key as well as the strange night sky the previous evening. It seemed that the sun was erupting. Astronomers had reported a series of unusual sunspots the previous few days, which had given rise to magnificent auroras while wreaking havoc with magnets. In New England people reported they were able to read newspapers by the night light. It was so bright green over the Rocky Mountains that gold miners got woken up in the middle of the night and were so confused they started making breakfast. In various areas people rushed to churches, convinced that the end of the world was at hand, although they also complimented the Lord on the magnificent colors of the end of the world. Telegraph operators across the country were struggling with traitorous keys that had decided to work or not work entirely of their own volition. Hitt silently chastised himself for doubting the telegrapher.

More pleasing to Hitt was a story that sleeping chairs had been installed in railway cars for the first time. Two nights earlier, passengers on the Chicago and Alton overnight between Bloomington and Chicago were able to stretch out and sleep as if they were in their own rocking beds.

Hitt suddenly realized he was lollygagging and strode briskly across the common to the courthouse. Fortunately, the spectators were still taking their seats when he arrived. The courtroom was even more densely packed than it had been the previous day. People had even taken seats on the windowsills, their legs hanging in midair. As he scanned the courtroom, he noticed several people who had not been present the previous day, among them two young boys and an elderly man with an antique ear horn. He went back to work, checking the same ink supply for the

ABRAHAM LINCOLN PRESIDENTIAL LIBRARY AND MUSEUM

The courthouse in which Lincoln's last murder trial took place was built for less than $10,000—although to make that possible the massive Ionic columns in front were actually made of wood and hollow inside. Lincoln was a familiar figure here and tried several hundred cases in this building.

fourth time. Suddenly the courtroom began buzzing, and he looked up from his business as Quinn Harrison and the defense team entered the courtroom. Lincoln took off his tall hat and began spreading his papers. As before, he removed his coat and hung it on the back of his chair, then rolled up his sleeves a workingman's distance. Logan straightened his stack on the table. Quinn Harrison, the natural excitement of the first day worn off, seemed completely lacking energy, as if he might fall to the floor in a pile if he had to stand again.

Judge Rice gaveled the court into session. Turning to the defense, he nodded for them to go ahead, reminding them, "Now it's your turn."

Logan stood and summoned Benjamin Short to the stand. Short had been properly named; he was small, stout and his jowls sagged as if they might be melting in this heat wave. Short acknowledged that he kept a drugstore in Pleasant Plains and was in that store on the morning of July 16. Logan asked formally, "Do you recollect anything of an affray that took place there in which Greek Crafton and Quinn Harrison took part?" As if it might be possible to have forgotten as much as a snail's nail about that fatal fight. Logan led him into his rambling narrative. "State when Quinn Harrison came into your store and what followed." Fully aware this would be dramatic testimony, he added, "Reflect and take it deliberately."

Short scratched his head, demonstrating his commitment to reflection, then recounted, "I think it was about the sixteenth of July at about half-past eight o'clock in the morning he came in. I think the mail had not come in. It was very nearly time for the mail, I remember, when Greek came in. Quinn Harrison was sitting on a stool near the door, I supposed four or five feet from the door inside I should judge. I was sitting side-by-side with him. We were looking over a newspaper, both over the same paper. We remained in that condition until Greek came in."

His testimony almost duplicated the story he had told at the first hearing. Logan urged him to continue, asking, "What then took place? State it all."

"Greek took hold of him the first thing I saw after he came in. He took hold of Quinn around the arms and the affray commenced. We rose up immediately, all three of us. Harrison and myself were sitting and rose up and I attempted to separate them. There seemed to be a fight. John Crafton came up. He had come into the store previously. He took hold and we went back together to the rear of the store all hands." As he told his story, he began speaking more rapidly; his voice also rose a pitch, and he began describing the event with his hands, waving them about in a growing frenzy. "John took hold of my left arm. I was endeavoring to separate the boys." That came almost as a plea for understanding. He tried to stop it. He tried.

"We all went back together and in going back Harrison got hold of the railing that goes around the counter and was pulled loose from that. He was pulled loose from it by all the other parties, I suppose. John was pulling on me; I was pulling on Greek and pulling on Quinn. He was pulled loose, I think. I think I saw a blow struck by Greek sometime during the affray. I think it was when we had got back to the rear—the back part of the building. I did not notice where he struck him. I did not see when Harrison struck. Greek was between Harrison and me. When John Crafton took hold of me he told me to 'Let him alone, Greek should whip him...'" He paused here, remembering now, his eyes focused on the past, then repeated but this time so softly, "'Greek should whip him.'"

Short looked up, directly at Logan, and continued,

"I don't remember anything said when Greek came in and took hold of Harrison." He paused again, and to his own surprise corrected himself. "Yes, Quinn told him he didn't want to fight him and wouldn't fight him—told him to keep off. That was after he took hold of him. Just as he took hold of him. As he took hold of him he told him to keep off, he didn't want to fight him. He still kept hold of him."

Logan made sure to emphasize his client's reluctance to fight, asking, "Did you see Quinn—" always Quinn, never Peachy or Mr. Harrison, an effort to humanize him without assuming too much familiarity and a subtle reminder to the jury that their client was well-known to many in the community "—do anything except take hold of the railing?"

"No, sir! I don't know the time the knife was stuck into Greek. I saw Greek striking at Quinn but I don't know where he was stuck. I looked towards the door. There were some men coming and I was anxious for them to get them separated and stopped."

"Did Quinn make any advance to him?" In other words, did Quinn take an offensive or defensive posture?

Short was clear, he made no threatening moves. "Greek took hold of him before he got up. He commenced raising just as Greek took hold of him. Greek went right to Harrison and took hold of him right straight. I think he didn't say anything." He shook his head. "I don't recollect. He had his coat off. I could not say whether he had his hat off. I didn't see him pull it off. It was possibly off when he came through the door." Then it came to him in a snap. "He pulled it off and threw it down and made right at Harrison before he began to rise. What Harrison said was just about the time he took hold of him—about his not wanting to fight him." Logan looked at Lincoln and sat down, while Lincoln stood.

He made a point of removing his glasses and placing them in front of him. "Mr. Short," he began, "where was John Crafton when the difficulty commenced with Quinn?"

"He was in the back part of the building. He had been there a very short time—I could not say just how long."

"He had been in more than once that morning?" The point Lincoln was trying to make was obvious; the Crafton brothers had been waiting for Harrison. Maybe even laying a trap for him.

"I don't know. I don't know what he was doing there."

"Did he say anything about having come because he expected somebody to leave money for him?"

"He did not to my recollection."

Lincoln paused, allowing that discrepancy to harden. Then just to make certain jurors understood, he asked with mock surprise, "He did not ask you if anybody had left money with you, and finally have some talk that it might have been left with Mr. Hart?"

Short was definite about it. "I don't recollect anything of it. I don't recollect any talk between him and me about money at all."

Hitt snuck a quick glance at the gallery. The spectators were fully absorbed in the questioning. He noted that Mr. Lincoln's entire demeanor was significantly different than the previous day; his voice was stronger and carried more inflection. He picked certain words out of each statement and hit them hard. "Which counter did Quinn hold on to by the rail?"

"The east counter on the east side of the room. When Greek entered, we were sitting with our backs to the counter on the west side. Quinn held on to the counter on…" Hitt completely missed the word, indicating that with a series of dashes. "—side of the pass way. It was an iron railing. Half an inch. He caught hold of that."

"You have not told what happened after Greek was cut?" Although this was a statement more than a question, Palmer did not object. This was considered entirely acceptable.

Short nodded. He knew what Lincoln was referring to. "After they were separated John Crafton threw some scales and some glasses and a stool at Harrison. John got cut somehow in the scrape. I did not see that. After they were separated and Quinn and Greek got apart Quinn kept pretty close to the rear of the store, I think in one corner near as I can recollect. He kept his ground and John threw at him. He asked if he had no friends here, I don't recollect whether he said it more than once but I recollect hearing him say that once. He asked if he had no friends there."

"Did he make at anybody after the separation from Greek?"

"No, I think not."

"Or at any other time?"

"No, sir, I think not. I didn't see him make at anybody during the whole affair."

Lincoln pushed harder at this theme—Harrison was always defending himself—asking, "As he went down did anyone push him?"

"They dragged him down," Short said, his whole manner an accusation. They, the Crafton brothers, working together, dragged him to the ground. "He seemed to be pulling the other way. I had hold of Greek and Greek had hold of Quinn."

"And by the general swing down that way, Quinn was dragged along down?"

"Yes, I think Quinn was pulled loose from the counter."

For the first time, Hitt saw Lincoln's lips quiver, as if he had started a smile but caught himself in time. Mr. Logan interrupted at that moment, also standing. "Mr. Short, when Quinn was holding on the railing was Greek behind him or facing him?"

"Facing him," Short said. Of that he was certain. "Greek had him around above the arms. I think he was very nearly that way when he first took hold of him. He took hold of Quinn from behind. Quinn took hold of me just as he took hold of him. His back was very nearly to him in the commencement. I did not see them face-to-face at any time during the scuffle. They might have been but I don't know whether they were or not. I didn't see them so. It was towards the rear of the store..." Again Hitt lost some words, but wrote down what he had heard. "...when my face was towards the door if their faces were ever together."

Lincoln asked a significant question that called for Mr. Short to state his opinion, as the fact could not be known. "Which was the larger and stronger man of the two?"

Mr. Palmer once again did not object, letting a witness have his say being the accepted practice. "I think Greek was the largest considerably if there was any difference. The strongest man also, I think."

Logan and Lincoln sat, satisfied. Palmer stood and acknowledged with a pleasant nod the fine job done by the opposition. As he did, Judge Rice shifted in his chair, searching for comfort. The heat was beginning to fill the courtroom, but there was nothing to be done about it. "Mr. Short," Palmer began, indicating the diagram of the store, "you observe an imperfect sort of figure here on the floor meant to represent your store. Describe it."

"Yes." He proceeded to repeat the explanation given previously, adding little to it beyond a guess that the store ran thirty feet deep. He also agreed he was about four or five feet from the front door when Greek entered. Palmer asked him what he was doing when Greek came in. "Looking over a newspaper," he said. "Greek's step on the door first attracted my attention. I suppose I stopped reading. I might have commenced again. It's a mere matter of impression."

Palmer had a calm and reassuring way about him as he asked his questions to the defense witness. There was no suggestion he was questioning Mr. Short's memory of the events, rather he was simply attempting to clarify them. "Was your attention attracted to Harrison as Greek came in?"

"Yes, when he took hold of him. I think not until then. He was sitting as close to me as he could when Greek took hold of him. I think both of us were leaning against the counter. It's about the usual height of counters. We were sitting on stools and leaning against the counter. When my attention was first attracted to Harrison, Crafton had not taken hold of him. I saw him before he took hold of him. I think Harrison was still leaning against the counter."

Palmer was dramatically confused, wondering, "Then how did Crafton get towards his back?"

"He turned his face towards me just as he took hold of him. He turned and his side towards me," Short said, twisting in demonstration. "He took hold of him just as he turned."

"Then he was not leaning against the counter when Crafton took hold of him?" Palmer was doing an admirable job poking holes in Short's testimony. Pinpricks, really, small inconsistencies, but together they weakened the fabric.

"It might have been touching him and it might not.

Harrison told him just about the time he took hold of him that he didn't want to fight him. I saw Crafton approaching. I discovered that he was approaching Harrison just as he came into the store. He went right towards Harrison."

"Did you rise?"

"No, sir."

"Did you apprehend any difficulty?"

"No, sir. Not until it commenced. In passing he passed along the usual passage way."

Outside, for the first time in several days, a cloud moving in front of the sun slightly darkened the courtroom. It wasn't nearly a promise of cooling rain, more a reminder that this heat was temporary. Palmer continued drawing his picture for the jury. "Then he was approaching you as well as Harrison when you first observed him?"

"Yes," Short answered, but with less conviction than previously, clearly wondering if Palmer was leading him into some clever trap.

"When did you make up your mind that there was going to be a fight?"

"As soon as I discovered him taking off his coat. It was just as he stepped into the house."

Palmer considered that long enough for the jurors to understand he was considering it. Then he asked, "Wasn't there plenty of time for you to have placed yourself between them?"

The sarcasm of Short's answer showed he did not appreciate the inference. "I suppose I might have put myself between them—if I had been in a pretty smart hurry." For the first time he looked to the gallery for support. The spectators were transfixed, no support came.

"Did he walk or rush in?"

"He walked in passing."

"Did Harrison observe him taking off his coat?"

Short pursed his lips, "I don't know. I don't think he said anything. I suppose the noise Crafton made drew his attention."

"Do you know any reason why it should not?" At that moment his purpose became clear to the courtroom; he wanted to make certain the jurors understood that Harrison was not surprised and taken. In fact, Harrison was aware of Crafton's presence and had sufficient opportunity to defend himself. Judge Rice leaned back; his morning mood already lightened. So much time in a courtroom is filled with nonsense, often his most difficult task was fighting to stay awake when overwhelmed by boredom. But watching smart lawyers do their business was still a treat for him. Lincoln, Logan and Palmer were putting on a fine show.

"I don't know. I am very apt to look up when I hear steps at the door."

As friendly as can be, Mr. Palmer asked the witness, "Is it not true that as Crafton came up Harrison rose to his feet?"

Short was confused, but tried sticking to his story.

"He rose to his feet when Crafton took hold of him. I don't think he did before."

At the defense table Lincoln leaned over and whispered in Harrison's ear. Harrison nodded vigorously. Hitt wondered if there really were necessary words spoken or if instead, this was some sort of distraction to break Palmer's rhythm.

"Did not Crafton take hold of him just about the time he was rising?"

"About that time, he did."

"Did you observe Crafton take hold of him as he was rising?"

Short paused once again. The timing was getting a little confusing. He tried to remember exactly how he had told it before, but he couldn't be certain. "I don't think I did," he said, the conviction now gone from his voice.

"Did you observe whether Harrison's hands were or were not about his bosom as he was rising?"

"I did not. I don't think I observed more than his turning his face from him. He had hold of him at that time, or he was taking hold of him about that time." He felt on firmer ground, repeating that this was about the time Harrison said loudly that he did not want to fight.

Palmer took that opportunity to remind the jurors of Crafton's threats. "Did he say to Crafton that if he laid hands on him he would kill him?"

"I think not," Short said, but Palmer had got his work done. That reminder could not be unheard. Short continued, "Nothing about killing or defending himself I think. I think he said he didn't want to fight or wouldn't fight to keep him off."

Palmer turned his back on the witness and faced the jury as he asked, "Was the tone of his voice loud or moderate?"

"I don't recollect."

"Did he curse Crafton?"

"He did not."

He turned around then and faced Short, asking, "That's all you recollect?"

"Yes."

Palmer then took Short through the fight again, questioning the smallest details. When Short said, "I expect I put both hands between them until John came up and took hold of one of my hands. Crafton had him drawn up close," Palmer wondered, "Then how did you get your hand between them?"

Short was forced to back up. "They were drawn up very close but not so that I could not get my hands between them, I suppose. Crafton didn't seem to be doing anything but holding Harrison."

Palmer asked question after question about pushing and direction and distance, east and west and front and back, enough to confuse even the most keen juror. Occasionally he tried a tricky question, for example, "Which hand did he strike the blow with when you had hold of him?"

Short was ready for him. "He didn't strike it then. He struck it in the rear of the store."

"And, is it true Crafton made an effort to strike and you had hold of his arm? Was it a free blow or a blow a man would strike having his arm confined?"

Again, Short was on the mark. "I could not say as to that. I have no recollection of the time or place when I let go his right arm."

Hitt's graphospasm made itself known as a throbbing ache in his thumb. He picked up a thinner pen, trying to move the resting point on that thumb enough to hold back the pain. That was the most he could do about it.

Palmer continued snapping questions at Short, leaving him as little space as possible to relax his thoughts. "Who was nearest Greek when he was in the corner where he was stabbed?"

"I suppose, John. We were all close together. I was not pushing him when he was stabbed. I don't know whether I had hold of him or not. I don't remember seeing him leaning over the boxes."

As Hitt took down the words, he wondered if the jurors also speculated on the quality of Short's memory. There seemed to be a lot of forgetting: *I could not say. I have no recollection. I don't know. I don't recollect. I don't remember.* The steno man supposed he should not be surprised that Short's memory, which had been so detailed only moments earlier under Lincoln and Logan, had lapsed dreadfully under Palmer's withering questioning.

Palmer's cross-examination was impressive, as he managed through his questions to raise doubt about the accuracy of Short's memory while suggesting quite a different interpretation of the facts. "Did you notice that Harrison had hold of Crafton at that time (of the striking)?"

"No, sir, I did not. I could not state their position when they got to the south counter because I was not looking that way." Short made it clear that he did not see Quinn stick Crafton, telling Palmer, "I was looking towards the door. I don't recollect the transaction after I broke loose. I have no very clear idea about it after that."

Palmer asked Short why he failed to stop the fighting.

"I suppose if I had laid out all my strength I might. I didn't try very hard. I thought we could get them to stop without it."

Palmer had moved his case forward small steps at a time. Lincoln and Logan had permitted that without interruption as it did not bear on their fundamental contention. But finally Palmer took a step into hostile territory, asking, "Did you see any sign of any particular danger to anybody or regard Harrison as in any danger of any immediate injury?"

For the first time, "Objection!" Lincoln and Logan rose as one. This went right to the heart of their case. If Harrison sensed no danger, he would not have reason to strike out with his knife. Neither lawyer stated the reason for their objection, as it was not necessary in that courtroom, but clearly this question called for the witness to state his own unqualified opinion. Most times opposing attorneys would allow it, but this went to the crux of their defense. It was a risk they could not take.

Judge Rice agreed. "Sustained."

Palmer asked a few more questions: Could Short have stopped the fight, did Short see the knife, did he know who owned the knife? Short could not know if he might have stopped it earlier, he saw the knife but "I did not examine it. I never had it in my hand."

Palmer gave the witness over to Lincoln to tidy up. "Do you know whether Quinn, when holding on the counter, held on with both hands or only one hand?"

"I think he held with both hands."

"I believe you said you had seen Quinn with that knife that day. When was it?"

"After the affair was over, or pretty well through. I think the first I saw of the knife was when the missiles were being thrown—when John was throwing. I never saw Quinn have a knife at any other time."

Lincoln smiled then. "Thank you, Mr. Short."

As Short stepped down, Robert Hitt quickly flexed his fingers and leaned back. Lincoln and Logan were huddled together in discussion. Judge Rice took a sip of water and stretched his arms behind his back as far as they might go, hoping for a bit of relief from the nagging lumbar pain.

Mr. Short walked through the courtroom with his head down, glancing at the jury. Hitt noted it often was possible to determine from that short walk how a witness believed he had fared in his testimony. A man holding his head high, looking at the gallery, perhaps even smiling, clearly believed he had done well for his side. A man walking with his head down, avoiding eye contact, felt poorly. Once, in an assault trial, Hitt had seen a witness whose testimony had been shredded pause and glare directly at the prosecutor; an obvious threat. But in this instance Mr. Short's walk seemed noncommittal.

Hitt wiped beads of sweat from his brow and while doing so chanced a look at Mr. Lincoln. He was not a man who allowed any show of emotion, but Hitt thought he detected the edges of his own satisfied smile.

CHAPTER ELEVEN

The defense called Dr. John Allen to the stand. Dr. Allen and Abe Lincoln had been long acquainted. The Reverend Allen had been one of Abe Lincoln's best friends in New Salem and had been among the small group of town elders who had urged the young man to try his hand at politics. Dr. Allen had organized and ran the first Sunday school in the village, which proved very popular, but also had founded the local Temperance Society, which was exceedingly less so.

In 1842 Lincoln had represented Dr. Allen, who had filed a claim for assault and battery against William P. Hill. The fight stemmed from a dispute over a fee owed to Dr. Allen. The larger man, Hill, had taken umbrage at Allen's slurs and attacked him, although little damage was done. Lincoln had successfully placed the blame on Hill, who was assessed $20 in real damages plus court costs.

Four years later, in late January 1846, Lincoln had been campaigning for Congress in Petersburg when a traveler from Aurora, Mr. H. C. Gibson, collapsed. Dr. Allen, who coincidentally also happened to be in that town, was called but together they could do nothing to save the man. After his death Lincoln and Allen jointly signed a receipt that they "have found in H. C. Gibsons Pokets ten dollars and fifteen cts in cash."

Since then they had seen each other only occasionally, but Dr. Allen had remained a strong supporter of Lincoln's ambitions. On the morning of July 16, Dr. Allen had business at Short's drugstore and arrived there after the brawling had begun. But what he saw that day was not the reason his testimony was vital to the defense. Stephen Logan asked him, "Did you see the whole or any part of the scrape between Quinn Harrison and Greek Crafton on the sixteenth of July?"

"I saw some of it," Dr. Allen said. "I didn't see the first commencement of it."

"State what you saw," Logan asked. "Did you go there with the expectation of seeing anything? Did you hear Greek Crafton make any threats before on that day?"

Palmer was up quick as a shot to object. When asked by Judge Rice the nature of this objection, Palmer said there should be no testimony allowed about threats supposedly made, "Unless evidence was shown in connection with it to bring the knowledge of these threats to the defendant before the affray."

Threats are made easily and often, Palmer contended; there may have been many threats made, but in this trial they had no relevance unless it could be proven that Quinn Harrison had direct knowledge of them. Jurors could not presume that he had knowledge of them; they could not make that connection through circumstance; it had to be shown with evidence that Harrison was directly and personally aware of the threats. Then and only then could it be argued that he got himself the knife with the white handle. They had reached the heart of the defense's argument.

Palmer was standing, Logan was standing. Lincoln stood. Judge Rice instructed the bailiff to remove the jury for this argument.

Hitt laid down his pen. The whole of this legal discussion was not to be transcribed, rather the gist of these arguments stated clearly and concisely. He watched the jury filing out, somewhat amazed how in such a brief time he had begun to consider these twelve individuals as a whole, as a jury. In a manner he could not describe, they seemed to have lost their individuality.

Palmer spoke first, making a formidable legal argument. The jury was being asked to consider the defendant's state of mind when he committed the stabbing and therefore was entitled to hear only those facts that might influence him. Whether or not Crafton made threats, and Palmer gave no concession that he did, was not relevant and would serve only to inflame the men of the jury without additional evidence that Harrison was aware of those words. Having prepared for this argument, he was armed with a concise history of hearsay evidence.

Among the first instances in which its admissibility was challenged was in the treason trial of Sir Walter Raleigh, in 1603. The telling evidence against him was the supposed confession of his alleged co-conspirator, who was being kept in the Tower of London. Raleigh contended that this man, Lord Cobham, had recanted and demanded that he be brought in to the courtroom so his own words might be heard. That request was denied and Raleigh was executed. The resultant outrage caused courts to begin reconsidering allowing hearsay evidence.

Palmer argued vigorously for several minutes pointing out the obvious: Greek Crafton could not be present in the courtroom to confirm or deny that he actually had spoken those words. And while he held great respect for Dr. Allen, the law was plain on the subject. He also stated that even if those words had been uttered, without Crafton's testimony there was no way to discern the true intent of his words. Precisely the same words might be spoken in all seriousness or in complete jest; they might be an affirmation or complete astonishment, depending on the intonation and punctuation. A phrase as innocuous as, *I did it*, takes on a completely different meaning if said as a statement: *I did it!* or asked as a question: *I did it?* The insertion of a pause might easily change the entire meaning of a sentence. It is not right, Palmer cautioned, to attribute a meaning to a dead man's words that might not be what was intended. Particularly on a point this significant.

It was a strong argument, Lincoln acknowledged as he commended Mr. Palmer for his thoughtful presentation, but he reckoned it had little bearing on this argument. It was not only Harrison's state of mind that Mr. Logan desired to demonstrate to the jury, but rather also that of Greek Crafton. Was it not necessary to determine Crafton's motive? Clearly, if Crafton started the fight with the intent to gravely injure Harrison, it made Harrison's need to resist that much more persuasive. And the only way to show that was through Crafton's own words.

Hitt was taken with the transformation in Lincoln's demeanor as he argued his points. He noticed a distinct change between the folksy, almost breezy Lincoln when the jury was present and the more formal attorney now presenting a legal argument. He was at home in the law, using the language of the courtroom with deftness and precision to bring the judge to his position. Rather than standing behind his table, he moved around the courtroom as he made his points. At first his hands were locked behind his back, but gradually he began to employ them, especially his right hand, poking at the air to give emphasis to a point. His voice never wavered, remaining firm and strong, and he commanded the courtroom so totally that it was not necessary to raise it.

Lincoln roamed the courtroom making his plea, reminding Judge Rice that the law might be flexible on such points, and with a young man's life in the balance it was far better to follow the spirit of the law rather than be held hostage to a strict interpretation. Let him have his best defense, Lincoln pleaded. As his arguments continued, Hitt was struck again, as he had been during the debates, by Lincoln's obvious respect for his words. People liked to call him plainspoken, but what Hitt noticed was the ability to string those ordinary words together in a way that made people pause and think about them in a different light. It wasn't the product of school learning; Hitt had been to fine schools and never heard another man able to turn words into feelings quite like that, but what it was exactly he couldn't identify by a name.

Lincoln's approach to the law was well-known. During the Douglas debates, Hitt had read up on him, every scrap he could find. He had seen a scribbled copy of a letter Lincoln had written to Justice of the Peace John King, who had been elected only a year earlier and had turned to him for advice on the administration of justice. The letter had been circulated by friends eager to push Lincoln's political prospects, and Hitt had been so taken with it he'd made a copy for himself. He'd figured it was pretty good advice: "Listen well to all the evidence," Lincoln had written, "stripping yourself of all prejudice, if any you have, and throwing away if you can all technical law knowledge, hear the lawyers make their arguments as patiently as you can, and after the evidence and the lawyers' arguments are through, then stop one moment and ask yourself: What is the justice in this case? And let that sense of justice be your decision."

If the courthouse truly be "dedicated to the cause of justice," he continued, then this must be the proper way to conduct the business of law.

Hitt watched with admiration for the man as he pleaded with Judge Rice to admit this philosophy into his courtroom. What he did not know, though, was that this was but a preparation for the more contentious argument to come.

Logan followed him, pursuing the legal argument with skill, telling Judge Rice that the defense intended to satisfy Palmer's objections by putting additional witnesses on the stand who would testify that Harrison had been made aware of the threats, and that Allen's testimony would be confirmed by those men. Judge Rice took pride in allowing defendants as much opportunity as was legal to make their strongest argument. To his surprise, he had enjoyed this debate; compared to the level of mendacious discourse he often had to struggle through, this was a fine legal argument. It reminded him of the reason he had chosen the law so many years earlier. And, for just a few moments, he had been so enmeshed in it that he forgot the maddening heat. Fortunately, Stephen Logan's promise had given him leave to come down in favor of Palmer without fear of cutting off a viable defense. He announced, "Evidence of threats by the deceased Crafton standing alone are inadmissible. There has to be evidence that the prisoner had knowledge of these threats before the fight and this testimony doesn't satisfy. Therefore the objection is sustained." Judge Rice then admitted his ambivalence on the issue, telling the defense that if there was a desire on their part to pursue this further, he would be pleased to reconsider his ruling and maybe, just maybe, if they could show him a different path, change it. Lincoln took his seat at the defense table, put on his glasses and busied himself reading some document while Logan responded. No doubt that feelings had been excited about this issue, Logan agreed, so rather than fighting it out all over again it probably would be best to proceed with other evidence and postpone further argument until the following morning.

Mr. Palmer consented to that solution; agreeing that Dr. Allen might be recalled to testify on this subject if the court reversed its decision. That settled it, at least temporarily. The jury was brought back in to the courtroom and when everyone was settled it was Lincoln who resumed the questioning. "Go on now," he asked Dr. Allen. "Tell the jury what you saw at the time of the altercation between Greek and Harrison."

Dr. Allen was a small, compact and neatly put together man. Everything about him seemed to be in order, although on a regular basis he would sweep his hand across his brow to push aside strands of hair that were no longer visible. In a great burst he responded, "Well, sir, when I first saw them together Greek Crafton had Quinn Harrison right around his arms and Mr. Short had hold of him and John was there by the side of Mr. Short and they were trying to pull Quinn from the counter and Quinn had hold of the guards of the counter. Greek had him around his arms and John hollered to Greek for 'to let him have it,' to 'give it to him' and then I saw Greek make a blow at Quinn.

"Then they got loose and as they got loose from the counter, Quinn pulled the counter toward the middle of the floor some two or three feet, dragged the counter. They staggered back towards the south end of the room and I didn't see what they did after that. Quinn hollered, 'For God's sake had he no friends there.'"

"Was that before or after he was dragged loose?"

"It was before *and* after," Dr. Allen said with certainty. "He hollered, 'Had he no friends there, for God's sake had he no friends there.'" He remained standing in the doorway, he continued, and from there, "I saw John throw the scales and strike Quinn in the side. He threw some glasses too..." Hitt paused here, unsure if Dr. Allen meant John had *also* thrown some glasses, or that he had thrown *two* glasses? It was one of those quirks of the English language that caused him anguish, but as he had been instructed, write down the word and let it be sorted out later. He guessed it was the number. "...two glasses...and a chair. It didn't hit him with the chair. I didn't see him hit him with the glasses. The scales did hit him."

Lincoln clarified a few issues then gave the witness to Palmer. The prosecutor took Dr. Allen through a description of the counter, it was of wood, and the piece that Quinn dragged was about ten feet long. Then he asked about the cutting. "Did you see Harrison strike him with a knife?"

He did not.

"Did you see a knife in Harrison's hands?"

He did not. "I never saw a knife that day," he said. But he did hear John Crafton say loudly, "'Let him have it,' and 'Give it to him.'" He heard him say it a few times, "as quick as he could," although he could not remember how often he had repeated those words.

As Dr. Allen stepped down, Skinny Thomas White followed him onto the stand. Skinny Tom was no kin to Fat Thomas White, although having been brought together by their shared name they had become friends. This was Skinny Tom, the farmer's son. Skinny Tom was nervous about testifying. For this appearance in the courtroom he had put on his best clothes and even shaved the peach fuzz off his face.

As he walked to the stand, there was some minor shuffling in the gallery as people stood up to move about in their place. Stephen Logan was going to use White to prove that Harrison knew of Greek's threats. He began by establishing that Skinny Tom had been at the picnic, then asked if he had spoken with Greek Crafton. "Yes, sir," Skinny Tom replied. "I heard him talking with another man and had some conversation with him myself."

Go ahead, Logan indicated with a sweep of his hand, *tell the story.*

"He was talking to a man, Mr. Purvines and was telling him he was going to whip Quinn Harrison if he came there that day to the picnic… Mr. Purvines told him he must not do it, it was not the place to have difficulties. It was no place there for it."

Without even a glance at Judge Rice, Logan asked politely, "Did you tell that to Quinn Harrison?"

Skinny Tom nodded firmly. "I told Quinn Harrison I heard a man say he was going to whip him there that day. I don't know as I told him it was Mr. Crafton, but I knew he knew it was Crafton soon as I spoke."

Logan spoke in a calming monotone, making it more of a friendly conversation than a life and death matter. "Was what you have related all that was said?"

Gaining comfort, to the amusement of those in the gallery who knew him, Skinny Tom became downright chatty. "Greek turned and asked Purvines if he was a friend of him. Says he, 'Are you a friend of mine?' Mr. Purvines says, 'Yes, I am your friend.' Says Greek to Mr. Purvines, 'I intend to whip Quinn Harrison for I have it from Wiley Crafton to do it.' He said Harrison was carrying six pistols and bowie knives for him. Then he remarked, that he had used these fellows…" Hitt noted that the witness was "rubbing his fists together."

"…for six months every night in Springfield."

As the scribe had learned years earlier, there were words and phrases spoken in every trial that made little sense to him. But his job was to record those words as they were spoken. This was one of those times.

Lincoln waited a moment then asked, "Who is Wiley Crafton that he said he had it from?"

The answer seemed obvious to White, though he hadn't thought to tell the jury. "His father is the only Wiley Crafton that I know."

"Did he say anything about six pistols?" Lincoln asked, referring to the weapons that he believed Harrison carried.

"I think he did. He said there was six pistols difference between them."

Lincoln saw the opening and barged through it. "Do you know anything about the relative size and strength of Greek Crafton and Quinn Harrison?" As he asked his question, he turned and indicated his client, Harrison, so the jury could have a good look at him while Skinny Tom answered.

"Yes, sir. Greek was the stoutest man. Greek was a very stout man to his size. I think he was probably 20-pounds heavier than Harrison. He was a very stout man for his size." Lincoln turned from his witness, perhaps to hide his amusement. When the tittering had concluded, he asked Skinny Tom additional questions about Harrison's state of being. "I cannot say much about him except by his looks. He has been sickly, I suppose."

"Have you known him to be in good health for the last three years?"

Skinny Tom shook his head firmly. "I can't answer. I have not been with him a great deal."

"Thank you, Mr. White," Lincoln concluded, turning to John Palmer and indicating with a wave of his hand that he might ask his questions.

John Palmer drew from the witness the fact that Crafton weighed no more than 145 pounds, although Skinny Tom continued to insist: "He was a stout man to his size." His use of the word *stout* resulted an echo of the earlier laughter. But he went a mite too far when he asked about Harrison's illness, wondering, "What was the matter? Did he have the ague? Did he have fever? Did you see him shaking?"

Judge Rice asked Skinny Tom if he needed to hear the question again. He did not. "I can't tell what ails him," Skinny Tom said, "I've seen him look very feeble."

With that answer Palmer let loose of the witness. As he left the courtroom a few of his pals gave him a hearty slap on his back and told him that he'd done good.

His testimony had served the defense well; leaving the impression of a smaller man being bullied by a larger, stronger man had made a telling emotional argument. Good people understood that sometimes a man could be pushed to the end of his rope. Nobody much objected to a fair fight, but it also was accepted that a man had a God-given right to prevent himself from being pummeled by a much bigger man.

To reinforce this notion the defense called Dr. Albert Atherton, the young physician who had taken over Dr. Stone's practice. As in many growing locales, one doctor was favored by the working classes while another might minister mostly to the more successful citizens. The handsome Dr. Atherton was the delight of the wealthy citizens of Springfield and its surroundings, among them the Harrison family.

In trials held years later, so-called expert witnesses would find it absolutely necessary to establish their credentials before being permitted to testify: Where did you go to medical school, Doctor? How long have you been practicing medicine, Doctor? Would you consider yourself expert on physical strength? How often have you testified as to the strength or lack of strength of a patient? A whole body of law would eventually evolve as to what constitutes expert testimony. But not yet, not at this point. Most of the people in Springfield knew that Dr. Atherton took care of the rich families and that seemed to be good enough for them. He had to know what he was doing or those folks wouldn't be using him. Albert Atherton had settled in Sangamon County six years earlier, soon after finishing his studies, and right away made himself part of the community by buying 160 acres outside Pleasant Plains to farm. A few years later he opened a grocery and drugstore in town. Far as anyone in that courtroom was concerned, he had proved long ago that his word was to be respected.

Dr. Atherton began his testimony by explaining he had been Peyton Harrison's family doctor for "some five or six years." Yes, he responded to Logan, he had been taking care of the son, Quinn Harrison, and "For the last three years I have taken him to be a man of feeble health and strength."

Logan continued to work through Dr. Atherton to create a sorry picture of the accused, asking, "Has it been advisable from the state of his health and prudent for him to do any work?"

Being too weak to work was a concept understood, and feared, by the men of Sangamon County. A man had to work to survive; everyone knew that. When you no longer could work, you clearly weren't too far from death's door. Dr. Atherton supported that conclusion. "I think, sir, he has not been able to make a hand at any ordinary labor..." Hitt dipped his pen and came up dry. He quickly switched inkwells but lost a few words in the process. "...probably able to do a part of a man's work for at least the past two years."

As this line of questioning continued, it became apparent to Hitt that just about everyone in the courtroom was stealing a glance or two at the defendant. Quinn Harrison had no doubt been made ready for this, as he sat calmly in his seat his eyes barrel-locked on Logan.

Although Hitt had worked in only a limited number of trials, he had seen defendants respond in different and often telling ways. Some of them were animated, talking endlessly with their lawyers while others sat sullen and defeated from the first. One sad man facing years behind bars put his sanity into question by laughing loudly at inappropriate moments. Harrison's calmness was to be expected, but it was his physical presence that was at issue. To the steno man, the defendant looked slight, his shirt sagging on his shoulders, his exposed arms thin without much indication of biceps, his complexion drawn. There was no suggestion of any type of strength—physical, mental or emotional—in his appearance. He simply seemed to be there, nothing about him in any way memorable.

Several members of the jury remained focused on Harrison as Logan continued to question Dr. Atherton; but if they were waiting for a telltale emotional response they were sorely disappointed.

"Do you recollect to have examined the state of his health for any particular purpose, for road labor or anything of that sort?"

"I did not consider him at the time able to perform labor on the road. Just at that time."

After some quibbles about his general condition, Logan asked a question that the defense believed would resonate with the jury.

"From the state of his health for the past three years, would you think that he has been or is in such a state of health that he may be expected to have the strength of a man of his size usually has?"

Dr. Atherton gave Logan the answer he needed. "No, sir, I don't think he has the strength that his appearance would seem to indicate."

The relationship between size and strength wasn't quite as fixed then as it would become. The fact that looks can be mighty deceiving had been brought home only two months earlier. The *Chicago Tribune* had created quite a stir by reporting that twenty-five-year-old Dr. George Winship, whose prodigious feats of strength had allowed him to proclaim that he was "the strongest man in the world," had swooned twice while delivering a lecture on "Physical Education" in Boston. Although only five foot seven and 143 pounds, ironically almost exactly the size given for Greek Crafton, "Winship can raise a barrel of flour from the floor onto his shoulder, can raise himself with either little finger, until his chin is half a foot above it, can raise 200-pounds with either little finger and can lift with the hands 926 pounds deadweight—Topham, the strong man of England, could only raise 800 lbs in the same way." And yet a speech had defeated him.

The prosecutor might also have pointed out to the jury that his adversary, Lincoln, was hardly a muscleman yet was renowned for his feats of strength. Instead John Palmer stood up and pointedly removed his own jacket, revealing a well-proportioned torso, then asked Dr. Atherton, "Have you drawn your inference except from his appearance?"

Dr. Atherton seemed to interpret the question as an affront, as if he didn't feel it polite to question his opinion. "He has occasionally been unwell," he responded, "and I have been called upon to attend him."

In the years he'd been the family doctor, Palmer asked, exactly how often had he treated Quinn Harrison? "I don't recollect," Dr. Atherton admitted, but did his best by adding, "Several times. I was called to treat him first for pneumonia, or inflammation of the lungs, that was probably four years ago. He has had several attacks of the same since that one. I don't recollect what was matter with him at the time he was unable to do road labor. He was not quite so well then as usual, I think."

That was Palmer's last question. Dr. Atherton stood to his full height and walked proudly out of the courtroom, without the slightest reference to Mr. Palmer.

There was a stirring in the crowd as people took advantage of the break between witnesses to stretch their legs, to stand up and jangle, even expectorating their chaw and replacing it. Several people dipped their pocket squares into a bucket of water provided for that purpose and mopped themselves cool.

They had just about settled back down when Abe Lincoln called the Reverend Peter Cartwright to the stand. This was the witness the spectators had been anticipating. The animosity between Lincoln and Cartwright was well-known, and given the importance of the Reverend's testimony, people were more than curious to see how the two men got on.

The seventy-four-year-old minister walked with the aid of a lion's-head cane to the stand. He was dressed appropriately to remind the jury of his stature, wearing a black jacket, black vest, white shirt and the unmistakable wide black band of his office around his neck, covered in part by a white collar. Lincoln sat ramrod straight as Cartwright walked by him and took the stand. The two men did not look at each other. Some isolated applause threatened to grow into something bigger but was stopped quickly by Judge Rice. While it wasn't often the judge had a real celebrity in his courtroom, he wasn't about to change the decorum. "None of that in my courtroom, please. This is a place of justice, not a meeting, and we'll have none of that." Then he glanced at the witness and acknowledged pleasantly, "Reverend."

"Judge," Cartwright responded pleasantly.

The anticipation in the courtroom somehow managed to cut through the heat. Lincoln and Cartwright had been going at each other for twenty-seven years with neither one of them securing victory. The insults between them had at times been pretty thick. Cartwright had once commented that Lincoln was so angular that "if you should drop a plummet from the center of his head it would cut him three times before it touched his feet."

During a debate, Lincoln suggested that the Reverend Cartwright's inability to state a firm position on an important political matter reminded him of a hunter who complained that deer were red in the summer and gray in the winter, so that at times a deer might resemble a calf. The hunter had managed to bag a deer from so far away that, he said, "I shot at it so as to hit it if it was a deer and miss it if it was a calf."

For almost three decades the two men had found precious little in common; from the propriety of bringing religion into the political discourse to their views on how to deal with slavery. They had argued and belittled each other through two campaigns in which they stood in opposition for election, as well as many more in which they supported different candidates.

But finally, at this late stage, they had found a common cause. The fate of Peachy Quinn Harrison had brought them together. Like everyone else in the courtroom, Hitt leaned back in his seat and looked from one to the other. The Reverend Cartwright seemed to struggle to make himself comfortable in the hard wooden seat, looking down as he tried to move it to a slightly different position. Lincoln, too, avoided looking at his longtime adversary, busying himself leafing through papers that he either dropped into his hat or piled on his desk. Logan sat looking at the Reverend Cartwright, an amused smile on his face as he watched these two men doing everything possible to avoid being caught with the first glance.

Stephen Logan had known both men for decades. He liked them both, and knew from his own experience where they were hardest to bear. For him, this was important knowledge. The trial might well be won or lost in the next few moments, and his ability to attack those weak points could prove to be the difference.

Judge Rice was certainly as intrigued as everyone else by this unusual situation. He could not remember another instance of lifelong adversaries winding up in the same pot. He took a sip of water and slightly adjusted his position. So much of a life spent on the bench was repetitive and admittedly there were times every judge struggled to stay awake. But not at this time—this was a confrontation of giants.

CHAPTER TWELVE

While the Reverend Peter Cartwright remained a robust and imposing figure in his seventy-fifth year, the challenge of bringing the Lord's word to the forests and prairies had reduced his once resonant voice to a fading echo of what it once had been. The man whose thunderous sermons once "shook the heavens" now was forced to speak as loudly as he was able just for his scratchy words to reach the back of the courtroom. But those same years had not at all diminished his intensity; his passion for bringing the word of the Lord to his flock was still on display. No one in the courtroom could miss it.

"Uncle Peter" Cartwright was well loved, and

greatly feared. No one would dare estimate the number of souls Peter Cartwright had saved in his long mission, but his objective in this courtroom was easily calculated; he was there to protect his grandson by telling the truth of what he had heard from Greek Crafton. His testimony, everyone knew, would be blunt. Cartwright was not known to hold back on his words. Once, for example, he had told Old Hickory, General Andrew Jackson, right to his face that he held no special place in the Lord's heart, and that he would be damned to hell quick as any other sinner who did not repent. The delighted general had responded with the wish that he had a few thousand officers as tough as Peter Cartwright.

Stephen Logan had spent substantial time with him before the inquest. While Hitt did not know for certain, he speculated that Cartwright and Lincoln must have met to discuss this testimony. It would have been an uncomfortable meeting for both men, but it was just too important. Undoubtedly, they would have been polite, putting aside the differences that had existed between them since Cartwright had preached regularly in New Salem. Even back then Lincoln had found it unseemly that Cartwright used his pulpit to further his political goals. But here they needed each other if both were to achieve their mutual aim.

Hitt imagined the meeting as brief and formal; Logan undoubtedly explaining to Cartwright the difficulties they faced in getting Judge Rice to allow his testimony, then Lincoln might have taken him through the basic issues point by point. Cartwright needn't remind those jurors of his distinguished—Hitt doubted that Lincoln would have used that word—his *notable* career and achievements as he carried those with him in his person. Lincoln was known to advise his witnesses to be truthful and be brief, suggestions that Cartwright at other times might have found insulting—he was always truthful and no one had the right to limit his speech. Hitt envisioned that at the end of the meeting they might have shaken hands—and later both of them would have washed well.

After much deliberation and discussion, Logan had decided it was better that he, not Lincoln, conduct the examination and Lincoln agreed. So to the disappointment of the spectators who had been awaiting this complicated collaboration between two great men, it was Logan rather than Lincoln who stood to begin the questioning: "Have you, Dr. Cartwright, been pretty well acquainted in the family of Mr. Harrison, particularly with Quinn?"

There was nary a sound in the courtroom as he responded.

Everyone knew the stakes here. "Yes, sir. Ever since he was an infant."

"State the reason of your being so particularly acquainted." Better to get this relationship out full-blown than pretend it was less than it was.

"Well, sir, he is my grandson, my eldest daughter's son."

"You have been about the house a good deal, Mr.

Cartwright?" Hitt thought it odd that Logan addressed the witness as *Mr.* Cartwright rather than *Reverend*, but recorded the words precisely as they were spoken.

"Oh yes, and he has been about my house frequently, often."

Logan stayed in place behind his table, ensuring that Cartwright had the full attention of the jurors. "Do you know anything about his health for several years past?"

"Yes, sir. I think I know very well."

Lincoln, fully aware that spectators might be watching him to gauge his reaction, remained as still as possible, his clasped hands resting in front of him on the table. Logan asked Cartwright to describe his grandson's ill health, "in your own way."

After clearing his throat, Cartwright said, "Well, he was..." He paused and appealed to Mr. Logan, "If I commence wrong set me right. Quinn was a very healthy child, as I understand it, for a good many years. He was an active, busy boy, surprisingly so for his age. But his health began to decline and I considered he was employed in too much business for his health, and I remonstrated with his father on the subject. His father had perceived it himself, and he relaxed from what he had formerly required of him. I know of my own certain knowledge, that for a good many years he has been a sickly puny boy, not able to do what we would call common manual labor."

Logan continued to make his point, asking, "Was there anything that called your attention to the propriety of his doing any work in which the state of his health was at risk?"

So long as the subject remained Cartwright's direct knowledge of his grandson's physical health, Palmer was content to remain silent. "Well, I considered that it was improper for him to do ordinary work in the business of the farm."

"That was on account of the state of his health?"

The preacher cleared his throat once again. "Yes, sir, his health has been very feeble, often laid up, and sometimes in the course of that time he was sick nigh unto death."

And then, without changing his tone at all, Logan changed the subject, asking Cartwright to "State whether you were with Greek Crafton, shortly before he died, and at the time he was expecting death, and if so, state what you heard, if you heard anything."

That brought John Palmer to his feet, where he objected loudly to the question. "Dying declarations are not admissible evidence, your Honor, and Mr. Logan is well aware of that."

Logan refuted that, pointing out that given specific circumstances, the confessions of men on their deathbed had often been deemed admissible.

Lincoln remained silent and seated as Logan and Palmer made their respective points. Judge Rice found no simple answer in the law and decided that in order for him to determine if the testimony could be allowed he needed to hear it in its entirety. To do that without influencing the jurors he had to remove them while the preacher was being questioned. After hearing it, he would determine whether to allow this part of Cartwright's testimony to be heard by the jury.

Now they were debating the law, and so Lincoln took the lead. Although he had great respect for the bench, Lincoln didn't understand why Judge Rice needed to know precisely what was said to determine if a man's dying declaration could be heard. After all, there already was plenty of good law on the subject.

The foundation of the law was its predictability, so before they went any further Lincoln wanted to know what standard the judge intended to apply when making his decision. What words would make it too strong for the jury to hear? It was a classic Lincoln legal strategy; make an argument in the form of a question.

Judge Rice clearly did not want to allow the jury to hear a dying declaration before determining its admissibility. If they proceeded as Mr. Lincoln seemed to be suggesting and the testimony turned out to be tainted, he would have to declare a mistrial and start the entire proceedings again. He was quite certain Mr. Lincoln didn't want that any more than he did, so the prudent course was to air it all out; then he would hear the arguments for and against and make his decision as to its admissibility. Years later the issue of admissibility likely would have been litigated long before Cartwright was called to the stand, but the standards that would eventually become common in trials were still evolving.

As long as the jury was going to go out, Judge Rice

decided to break for lunch. He ordered the bailiff to take the jurors back to the apartment and make sure they all got something good to eat. Then he banged hard twice on the sound block and said, "Let's go get us sumthin' to eat."

Hitt wiped his pen, laid it down, then stretched. Several members of the gallery made their way to the door, racing to go out back to the privy, but more of them remained in their places, unwilling to risk losing their seats for the afternoon session. Once again Hitt began his perambulations, nodding pleasantly to the boys as he passed them. This time they nodded back and said nothing, but rather seemed to shrink away.

Courtrooms are often quiet but when the Reverend Cartwright was called to the stand as the session resumed, the only sounds were of bodies shifting in the gallery, as spectators attempted to find a slightly better angle to view the witness. The courtroom felt odd, strangely out of sorts with the jury box sitting empty. To the surprise of some it was Logan who continued the questioning, urging, "Go on, Mr. Cartwright, and tell the story."

Cartwright nodded, and began, "I understand that this affray took place July 16. I was not at home, nor present at the time." He looked at Mr. Palmer and said, "I know nothing of the quarrel, pro or con, except by hearsay. I was friendly to all the parties, in good friendship and fellowship as far as I know." He returned his attention to Logan. "I passed through the village where this affray took place on my return

ABRAHAM LINCOLN PRESIDENTIAL LIBRARY AND MUSEUM

When the Reverend Peter Cartwright, a Methodist and revivalist, met Lincoln in the courtroom he was in the twilight of his career, while Lincoln's was in its ascendancy. At that time Cartwright was beloved and respected nationally, while Lincoln was still an enigma. Arguably he was more antislavery than Lincoln, believing the system to be immoral. After more than two decades as bitter enemies, this trial brought them to the same side.

from some lower counties home, obtained my papers and letters from the post office and heard not a word about this affray until I got home. When I entered the room, my wife asked me if I had heard of the dreadful affair. I told her, 'No, not a word.' When she repeated it, it shocked me exceedingly."

For the first time his gaze slid just a bit and appeared to fix on his grandson, and Lincoln. He cleared his throat then continued, "A day or two before this my horse had fallen upon me, and crippled me in my right hip. The day was exceedingly hot, and I was exceedingly faint, and sick. In a few minutes a messenger from the village, Mr. James Fanes, came in, saying that Greek Crafton, who was wounded, had sent for me to pray for him.

"I did not know how I was to get there. I had turned my horse into the pasture and I had no one to catch him, but he said he had brought a wagon and I could ride up. I thought, at first, I could not go, I was faint and weary: but I concluded to make the effort, and went.

"When I came to the village, I learned that he was at the house of his brother-in-law, Dr. Million. I made my way to the house and went into the room. It was considerably crowded and all present were affected; I was affected myself, deeply and solemnly too, for I had heard that it was supposed that the wound was mortal."

As the Reverend Cartwright settled into his story, his voice seemed to gain strength and timbre, but his eyes seemed slowly to lose their focus on the courtroom and drift into the past. "The wounded young man, Greek Crafton was lying on a little trundle bed that was near the floor, and I, not being able to stoop on account of the injury to my hip, I knelt down by his bedside and took him by the hand and expressed my deep regret and sorrow that this fatal calamity had fallen upon him—for I looked upon it as fatal from what I heard on all sides. I looked upon him as in a dying condition and felt accordingly. When I took him by the hand and thus expressed my deep sorrow he said, I think at that moment, 'The honest hour has come, and in a few moments I expect to stand before my final judge; do you think there are any mercy for me? Will you pray for me?'" Palmer sat back and let Cartwright have his complete say.

With the jury out of the courtroom there was no damage to be done to his case by this testimony and even he was touched by the aching in Cartwright's words.

"I then repeated my regret and sorrow at the circumstance, the fatal stab that I supposed he had received. I expressed my deep and heartfelt regret that this calamity had fallen upon him. He paused a moment and made this reply, I think literally, 'Yes,' said he, 'I have brought it upon myself, and I forgive Quinn,' or 'Quinn Harrison.' I am not certain which, he seemed as calm and composed as any man could possibly have been in his situation, and I really wondered that he could command himself as he did. He requested religious services, and we went through them as best we could over a dying man. I have visited hundreds of dying fellows and I thought I never saw a man in a dying condition more calm and composed, more perfectly than himself." Hitt could never recall ever being brought so strongly into a past moment by a witness. His craft had enabled him to perfect the art of dispassion, but here he saw the courage of the boy and felt it. He kept his head buried in his work so as not to reveal this professional lapse.

Cartwright shook his head in sadness as he once

again found himself at Greek's bedside. "His mind seemed to be entirely fixed on his final destiny and not on the sympathetic wailings of his friends around him. I then arose, being in extreme pain from my crippled hip, and it being extremely warm, stepped back two or three feet, not very distant. There were two doors to the room, to the north and south, and on the east and west windows that were hoisted. I stepped back to get the air. A lady handed me a seat; I sat down to get the benefit of the air, to relieve myself. Then I heard him repeat distinctively what he had said to me, 'I have brought it upon myself. I forgive Quinn and I want it said to all my friends that I have no enmity in my heart against any man, and if I die…'

"Previous to this, at the close of the services, he had said that he hoped for mercy from a final judge. '… If I die, I want it declared to all that I die in peace with God and all mankind.' I sat there in my pain. I cannot tell how long. I think I got through between one and two o'clock. I am not certain as to the time. I lingered there two or three hours. The room being crowded and the doctor not able under the circumstances to keep it clear for a free circulation of air, I requested the doctor to do so, for that was enough to kill the dying man, if nothing else. I then removed outside to the shade, for the sun was getting pretty low and soon returned home. That is the substance of what I can recollect."

He settled back in the seat, done. The courtroom remained silent. He took a kerchief from an inner pocket and wiped his brow. Then he looked up at Judge Rice, as if to ask if more was needed. Judge Rice looked to Logan and Lincoln to see if they had further questions. They did not. "Mr. Palmer?" he asked.

Palmer shook his head. "No, your Honor."

Judge Rice said pleasantly, "Thank you, Pastor, thank you very much. Please stay around here a bit."

The courtroom breathed a collective sigh, all the restrained noise letting loose at once. Seemed like everybody had something to say about that testimony. Judge Rice let it go for a bit, being busy straightening up his bench, while really considering his decision. Finally he picked up his gavel. It took him four hits to silence the room. "Quiet down!" he yelled. "Let's quiet it down."

Kidd, too, tried to clamp down the noise, shouting, "Quiet down, just quiet down."

Judge Rice sat fidgeting on the bench, scratching his head as he decided how to proceed. Hitt looked at Quinn Harrison, who sat rock-still at the table, his eyes locked on to the judge.

Judge Rice cleared his throat. He knew the stakes here, he said, and he wanted to make sure the defendant got a fair shake. Then he began to speak about the law, saying it was set down by men to bring order to a chaotic world. It occurred to Hitt that the judge was fumbling his way through his own mind to reach a decision. The judge continued, pointing out that since it had been made by man, then every single soul in the courtroom knew for dang sure it wasn't going to be perfect. During lunch, he said, he'd been giving this issue a lot of thought. In the past, whenever he'd found himself in a quandary, he'd discovered that Mr. Shakespeare usually had something smart to say that helped him sort it out. He picked up a sheet of paper and began reading, "'Between two hawks, which flies the higher pitch; Between two dogs, which hath the deeper mouth; Between two blades, which bears the better temper: Between two horses, which doth bear him best; Between two girls, which hath the merriest eye; I have perhaps some shallow spirit of judgment; But in these nice sharp quillets of the law, Good faith…'" He paused here for effect. "'I am no wiser than a daw.' No wiser than a daw."

There was one more quote he wanted to add, he said, that he thought pretty much everybody would agree with, and this one came from the British author Charles Dickens, who had written bluntly only two decades earlier in *Oliver Twist*: "The law is a ass, a idiot." That drew agreeable laughter from the spectators. "So what am I going to do here," Judge Rice wondered aloud.

Then he began talking about the law, explaining that where dying declarations were concerned, previous courts had seemed to find a significant difference between statements of fact and statements of opinion. Exceptions to the hearsay rule had most often been acceptable when a dying person had provided important information about the cause of their demise. Who done it, for example, and how it had been done.

But that was not the situation here, he said. There wasn't much dispute about that. And try as he might, he didn't see how this qualified...

A sudden motion to his left startled Hitt. With no warning, Lincoln erupted, springing from his chair and demanding, in a massive voice that rattled the courtroom walls, "Your honor! We need to see this through. Every last bit of it!"

Hitt was stunned. He had never seen such an unexpected explosion of passion in his life. Lincoln clearly had heard Judge Rice drifting toward ruling the evidence out and determined to stop him right quick, stop him cold before he reached that end. He was at the bench in the wink of an eye. The calm demeanor with which Lincoln had previously conducted himself was gone, replaced by a controlled fury. This ruling could not be allowed to stand. The words seemed to burst out of him; the law was not a game to be played, he thundered, as he roamed the courtroom, his unleashed energy making it impossible for him to remain still. Justitia, Lady Justice, has carried her scale in one hand since the time of the Roman Emperors, he said, a reminder that all the evidence must be weighed. All of it! It is not the job of the court, he yelled, shaking his index finger to emphasize that point, not the job of the court to push down the scales on one side, but rather make certain every last bit of grain is counted. Then, and only then, will the balance of justice be done.

He pointed to the prosecution table. "Those are able

people sitting there, your honor," he said. He knew them all, he knew how good they were. The last thing they needed or wanted was the court's help to fix their case. They got their own things to say about this, so let it all be heard, every word of it, then let the jury give it the appropriate weight.

Later, Hitt would describe the scene to that Chicago reporter, "Lincoln sprang to his feet and he seemed to be going right over the top of the bench on top of the judge. I never saw a demonstration of power manifested in any human being in my life equal to that."

This was an aspect of Lincoln's character Hitt had not seen. It was stirring. He was forceful and determined, his usually measured words this time dripping with emotion. It was clear he was giving no quarter to Judge Rice; he was attacking with a heady brew of logic and law.

"It was," Captain Kidd would later recall, "a display of anger, the like of which I never saw exhibited by him before or after. He roared in the excess of his denunciation of the action of the court."

Days later, William Herndon described the scene: "Lincoln rose to read a few authorities in support of his position. In his comments he kept within the bounds of propriety just far enough to avoid a reprimand for contempt of court. He characterized the continued rulings against him as not only unjust but foolish; and, figuratively speaking, he pealed the Court from head to foot. I shall never forget the scene. Lincoln had the crowd, and a portion of the bar,...

with him. He was wrought up to the point of madness. (He was) mad all over...he was alternately furious and eloquent, pursuing the Court with broad facts and pointed inquiries in marked and rapid succession. (To make his point he told a well-known story.) In early days a party of men went out hunting for a wild boar. But the game came upon them unawares, and scampering away they all climbed the trees save one, who, seizing the animal by the ears, undertook to hold him, but despairing of success cried out to his companions in the trees, 'For God's sake, boys, come down and help me let go!'" Lincoln had a firm grasp of his case, Herndon remembered, but he was getting no help in making it from the court. And should he let go of it, even for a moment, the result would be disastrous. Lincoln went on this way for several minutes, at times, Hitt thought, coming within sight of contempt. But Judge Rice allowed him to continue, until he walked back to the defense table, saying clear enough for the entire gallery to hear him, that the deceased has a right to be heard.

When he was seated, Judge Rice glared at him. "You finished?"

"I am, your honor. Thank you."

Judge Rice turned to address the gallery. "As I was saying before Mr. Lincoln found it necessary to give us all the benefit of his learned opinion..." He then proceeded to explain that, as he was saying before being interrupted, he couldn't see how the Reverend Cartwright's testimony qualified under the dying declaration exception, but being no wiser than a jackdaw he also didn't see how he could exclude it fairly, especially since the prosecution was ready to offer powerful rebuttal witnesses. On balance, he said, looking at Lincoln with an almost sarcastic smile, he saw more damage potentially being done by excluding this testimony than allowing it. So for that reason he had decided to let the jury hear it.

Herndon believed the reason for this ruling was not so much based in the law as his partner had "So effectually badgered the judge that, strange as it may seem, he pretended to see the error in his former position, and finally reversed his decision in Lincoln's favor."

"Thank you, your honor," Lincoln said, but his words were lost. Once again, the courtroom erupted. Just about every spectator held their own opinion on this, and those who had been proved right took the time to crow about it. Lincoln cupped his hands around Quinn Harrison's ear and said something, to which Harrison nodded vigorously. For the first time in the whole trial, he was smiling.

Hitt was surprised, although of course he showed no part of that. Judge Rice had seemed well on his way to a decision, but seemingly had been turned around by Lincoln. Well, he wouldn't be the first one, the steno man thought. Of course, he had no way of knowing if Judge Rice had been persuaded by Lincoln to change his mind or if he was always going there, but the result was the same: the jury would be permitted to hear the victim apparently taking blame for his own demise, and excusing the accused.

Before recalling the jury, Judge Rice asked Palmer if he wanted to have a say. The prosecutor said he accepted the decision that Crafton's alleged deathbed statement could be heard by the jury but asked that Cartwright's testimony be confined strictly to his declarations as to the facts of the killing. Of course, that interpretation of the ruling would effectively gut Cartwright's testimony. He argued about the relevance of the testimony, claiming most of it was immaterial to the issues that had to be adjudicated. He couldn't recall any part of the criminal statutes that said it mattered a whit if the victim forgave the accused. Under the law, he said, the heart of the testimony should be ruled "incompetent," or inadmissible.

Stephen Logan spoke for the defense, arguing that this went directly to the state of Greek Crafton's mind when he started the fight. The jury had the right to know that Greek had set out to beat Quinn Harrison.

Abe Lincoln followed him; having regained his equanimity he spoke calmly, briefly and eloquently about the right of the accused to present his best defense, calling it fundamental to the concept of a trial by a jury of your peers.

Judge Rice overruled Mr. Palmer's comparatively mild objection, then told the bailiff to go fetch the jury. When the jury had been reseated, Judge Rice told them that he was going to let them hear what the Reverend Cartwright had to say and that they better pay close attention because there were still some parts of it that he might later decide they shouldn't consider in their deliberations.

The Reverend Cartwright sat back in the witness chair. He may well have believed the Lord had put him in this awkward position, having made him the confessor to the young man slain by his grandson. Years later *McClure's Magazine* would describe the tension in the courtroom, writing, "No face in Illinois was better known than his, no life had been spent on a more relentless attack on evil. Eccentric and aggressive as he was, he was honored far and wide; and when he arose in the witness stand, his white hair crowned with this cruel sorrow, the most indifferent spectator felt that this examination would be unbearable." This being such an unusual occurrence, no one knew quite how to proceed. Mr. Logan shrugged and told him he might just as well go ahead and tell the whole story over again. The Reverend chuckled to himself and admitted, "I don't know but these lawyers have got me so tangled up that I don't know which end of me is up, but I will try to be as collected as I still can. If I understand myself, I am now to repeat what I have said once?"

Logan agreed, "If you so recollect it, unless something else occurs to you."

John Palmer told him, "State what took place at Dr. Million's and state the time."

The Reverend Cartwright said, "I don't know how

to get at it without my whole statement. I'll make it brief."

Lincoln suggested, "Please go over it all again, Dr. Cartwright."

And so he did. The *Illinois Journal* noted that in telling his story while the jury was out, he had testified in his renowned "characteristic and forceful manner," but in front of the jury he spoke "in a voice tremulous with age and feeling." He told of passing through the village. His horse falling. Going to Greek's bedside. Recalling Greek's words… But then, as before, he stared into the past and repeated Greek Crafton's words with real pain in his voice. "Will you pray for me?"

"I have no enmity in my heart." And, finally, "I die in peace with God and all mankind."

When he finished this second telling, he coughed a few times to break the sad silence of the courtroom, then addressed Judge Rice, asking respectfully, "If I could be released altogether I would like it, for I have an appointment in Jacksonville tomorrow."

Judge Rice asked the prosecution if they had any questions for Pastor Cartwright before he released him, and there was some surprise in the courtroom when Palmer said he did not want to keep the witness. Even Hitt was a bit taken aback by that. No questions? The obvious inference was that the prosecution did not believe the Reverend Cartwright had damaged their case so badly that they would risk alienating the jury by attacking this respected, beloved figure.

Better to let him go, and use other witnesses to refute his testimony.

None of the lawyers objected to that request, so the Reverend Cartwright was discharged from further attendance. As he walked slowly across the courtroom, he cocked his head and caught Lincoln's gaze, and nodded with satisfaction.

When the trial resumed, Lincoln told Judge Rice that the defense wanted to call those witnesses he had offered yesterday, people who were going to testify that between the fourth and sixteenth of July, Greek Crafton had continued to make threats against Quinn Harrison. The last one being made less than a half hour before the fighting, Lincoln said, which made it probable that Crafton was in the same state of thinking when he encountered Harrison.

Norman Broadwell was on his feet, objecting, even before Lincoln finished his pleading. Judge Rice immediately realized that he was once again caught in a legal poser as Broadwell explained his objection; this already had been a long and legally complicated day, and now they verged on another complex and time-consuming question. Broadwell acknowledged that if the defense could show the accused had direct knowledge of these threats, then they could have reason to call these witnesses. But, he emphasized, they couldn't. And that wasn't their intent. Instead they were trying to obfuscate the truth of it, presenting a concoction of this and that, trying to confuse the jury and prejudice them against the victim. Why, he railed, it's no different than claiming you're making an apple pie, but you're using blueberries because you got no apples. You can call it anything you want, but it ain't an apple pie. The defense can't prove the accused knew about the threats so all this does is rile up the jury. And there is no legal justification for putting these people on the stand.

Logan and Lincoln together refuted that argument, stating once again that their purpose in calling these witnesses was to show that Greek Crafton set out to fight their client and was in such a state that there was no telling what he might do. That's in keeping with the law.

Robert Hitt thought Judge Rice looked miserable. His head was bowed, and he was regularly wiping away the sweat. But after hearing both sides make their cases, he decided that he was still going to allow the defense to put their people on the stand. Lincoln called twenty-four-year-old Madison Cartwright, the eldest of the Reverend Cartwright's eight children and therefore Quinn Harrison's uncle, to the stand. Madison Cartwright, who had recently married Nancy Purvines, had served as his father's assistant on the trail, helping bring the Savior to the frontier. He carried himself with aplomb as he took his place in the witness box. He wore a fine coat and a perfectly set string tie, and somehow showed no signs of feeling the heat. Lincoln asked him pleasantly, "Were you present by the side of Greek Crafton's bed just immediately preceding his death and at the time your father was having a conversation with him?"

He was, he said. "I heard him, at the time Father approached him and spoke to him and told him that this was a severe calamity that had come over him, or something to that effect. I heard Greek say this, 'I brought it upon myself.'"

"Thank you, Mr. Cartwright," Lincoln said. The young man sat still, surprised that more was not asked of him. When Judge Rice asked Mr. Palmer if the prosecution wanted at him, Cartwright turned and looked expectantly at that side, but Mr. Palmer waved off any interest. As the young man left the box, there seemed to be an air of disappointment in him that he had no further role in this trial.

Next called was P. M. Carter, one of that number of young Sangamon County boys who had grown up around each other, their friendships and loyalties flowing and shifting like the nearby river. Abe Lincoln questioned him, asking him to affirm that he was acquainted with Greek Crafton "in his lifetime."

Carter was known as a sharp tack, a man with a quick mind, a quick wit and quick words. He was the fellow the others turned to for advice. Indeed, he told Lincoln, he was an acquaintance of Greek, adding, "And also with Quinn Harrison."

Lincoln asked him to tell, in his own way, what he heard Greek Crafton say about Quinn Harrison. He began in somewhat of a monotone, "Either the first of the week upon which the affray occurred or the last of the week previous, I can't state the day, Greek was in Mr. Turley's store and I was there also." Hitt filled another sheet of paper and as he put it aside and began his seventy-sixth page he missed a few words. "He came…that Quinn Harrison was 'a… Knife,'" was the best he could do and for clarity Hitt noted that Carter "(ran his hand down his bosom)."

"'(D)own there for me,' and says 'He will be down there this evening and I am going to give him a chance to use them.' I then took Greek to the other side of the house, to the west counter, and told him that I advised him as a friend to let Quinn alone, and if he would let Quinn alone I would go to his tail that he would never interrupt him, that Quinn would not speak to him. But, says I, 'If you ever attack Quinn I believe he is determined to kill you as he told you, and if you take the advice of a friend you will let him alone.'

"He just raised up and says he, 'If he is here this evening he has got to try it on!' That is all he said to me at that time that I now recollect."

Without the slightest change in his tone, Lincoln asked the telling question, "Previous to the affray did you tell Quinn of that?"

Carter hesitated; a keen man, he understood the intent. "I cannot say that I did," he sighed, trying to dredge up a memory that would not come, "but it seems to run in my mind that I did. But I can't refresh my mind of any time or place."

Lincoln found that curious. "It seems to run in your

mind," he repeated, somewhat quizzically, "but you can't be certain about it?"

Reaching the truth required a nifty twist of the language. "Yes, I can't," he said.

Lincoln responded in kind. "You was not there on the sixteenth?"

"No, sir!"

John Palmer joined in, enjoying the confusion. "When was that?"

"On the first of the week that the affray occurred, or the last of the week before that."

It had already been a long afternoon for Judge Rice, who at that moment was not particularly appreciative of his robes. He needed to take a few minutes to refresh himself. "Let's take us a little break," he told the courtroom. "I don't want to see any of you folks taken by the heat."

CHAPTER THIRTEEN

Hitt took advantage of the brief pause to go out back. Indoor plumbing was coming someday, it was inevitable; the Tremont Hotel in Boston and New York's Astor House already offered it to guests, London's Crystal Palace had been offering public facilities for nearly a decade, but in Springfield folks still relied on water closets and outhouses. Fortunately, the line was not long, probably because few people fancied waiting in the boiling sunshine.

When the trial resumed, the defense called the Reverend John Slater, who ministered to Greek Crafton throughout most of his final ordeal. Lincoln asked him to tell the court, "What you heard Greek say, if anything, in the presence of Uncle Peter Cartwright, as to answering him."

"Well, sir," he began, "I was present when Mr. Cartwright came in and when he advanced to the bed in which Mr. Crafton lay, and remarked to him that he was sorry to find him in that condition, that it was a dreadful calamity. Mr. Crafton then remarked, 'It is,' and then he said, 'The honest hour is upon me. I am a dying man and in a few minutes or hours at most I shall have finished my earthly existence.' And he inquired of Mr. Cartwright whether there was mercy for him. Mr. Cartwright answered in the affirmative that there was. He then remarked that notwithstanding the nature of the calamity he was glad to know in this last extremity, in this state that he was able to say that he had not aught in his heart against any mortal—neither Quinn nor anybody else, and that he had brought this upon himself. He then remarked, 'I forgive Quinn.'"

Lincoln recognized the perfect place to stop. There was nothing that he could get from this witness that would be more helpful. So he thanked him, glanced at Mr. Palmer and said, "John."

"Thank you, Abe," Palmer said. "Dr. Slater, was that all he said as to Quinn's connection with his death?" The tone of his voice inferred he had some additional knowledge.

"There was another remark afterward," the Reverend Slater admitted, but he dismissed it as "something of the same import."

Clearly Palmer felt differently. “Did he undertake in that remark to speak of the manner in which he had been injured?”

“No, sir. There was no reference to it.”

Palmer persisted. He was looking for something more. “Have you now detailed all he said in reference to the manner of his death and Harrison?”

“All he said at that time. He remarked after the doctor stepped out of the room…”

“Dr. Cartwright?” he asked, an effort to clarify which “doctor” he meant.

“Yes.”

“Was it a continuation of the same conversation?”

Lincoln leaned behind Harrison and whispered something to Logan, who shook his head in response.

“It was not continued to Mr. Cartwright. He was not in the room…”

Palmer was unusually aggressive with this witness, cutting him off before he finished his thought. “Do you recollect his taking his seat in the room and the conversation continuing?” It was not clear to anyone in the room where he was leading.

“Yes. I recollect his making a remark to Dr. Cartwright, to my mind afterwards that day in the room with Dr. Cartwright. I understood the remark to be made in general, but I don’t know whether it was made to anyone distinctly or not.”

“State what he said in that continued conversation.”

There was a slight hesitation in the Reverend Slater’s response. “The remark was about this, or it was similar in nature..." He paused completely, then admitted, "Whether precisely the same language or not I can't say; that he had brought this upon himself and that he wished all the world to understand that he died in peace with God and man."

"Any other remarks?"

"I think not that I recollect."

"Do you recollect any further allusion to Harrison?"

"No, sir. No allusion to Mr. Harrison."

Palmer was letting the cat out of the bag. "Do you recollect any remark about his being held while Harrison stabbed him?"

Throughout the courtroom people suddenly sat taller. "No, sir," Slater said flatly.

"Do you remember Mr. Short's name being used?"

"No, sir, not at that time."

Palmer stood directly in front the witness. "I speak now of any portion of a conversation which was continued up without interruption until it was ended?"

"No, sir," Slater said again. "Not at that time."

Palmer remained standing, waiting to see if the witness would add to his recollection. Silence often makes witnesses uneasy and they rush to fill it with more words than they had intended. In this instance, however, the Reverend Slater remained quiet. Finally Palmer said, "Thank you" and turned away.

Tom Turley was next, the owner of Turley's store, which became an important venue both in the week before and the day of the encounter. Turley seemed made to be a shopkeeper; he was short and wide-spread, with the center of his head bald and clean. All that was missing to complete his appearance was his apron. Again, Lincoln spoke for the defense, asking if he knew the young men. "I have been acquainted with them some," he said. He lived in Pleasant Plains, where he kept the post office. Asked by Lincoln if he "heard Greek Crafton say anything about Quinn Harrison," he responded, "I heard him make threats twice previous to that time, after the fourth and previous to the sixteenth.

"The first was on the Tuesday, I think, before the injury. He remarked that he was going to whip Quinn Harrison on sight. That was all I believe I recollect of hearing at that time. That was in my store. The next threat, I think, I heard on Thursday. I heard him say he was going to knock Quinn Harrison down and stamp him in the face. He said he understood that Quinn was carrying a knife and a pistol, for him a bowie knife, and he said that there was the difference between them."

"Did you say anything back to him?"

"I told him he had better not do that, that he was a larger man than Quinn and it would be no honor to whip him. I told him at the same time that he might get hurt if he undertook it."

Lincoln had Turley repeat the fact that Greek was a larger and stronger man, then quoted him, "The words were that he would knock him down and stamp him in the face?"

"Yes." He nodded in affirmation, yes.

Lincoln offered his appreciation and took his seat. As Palmer began his questions, Lincoln made a show of putting on his glasses and reading a note Logan had written for him. Palmer asked Mr. Turley if Greek had given a reason for his anger. He had not, shaking his head as he said, "I don't know that there was any occasion for the remarks."

Palmer obviously thought that was strange, using his question to make that point for the jury. "You don't recollect what gave rise to the remarks? No allusion to Harrison or any difficulty with him or what he had done or anything about him? But he simply remarked, without any introduction, that he intended to knock Quinn Harrison down and stamp him in the face?"

"Yes, sir," Turley insisted.

"And the conversation ended there?"

"Yes. If there was anything more said I can't recollect it."

"Thank you," Palmer said politely, but turning his back to the witness with a show of disdain.

The trial had picked up rhythm as the defense built its case piece by piece. Dr. John Allen, who had not been allowed to testify about threats he might have heard during his previous appearance in the witness box, was recalled to the stand. Judge Rice reminded him that his oath still counted.

"Nice to see you ag'in." Lincoln said, finally able to ask the witness, "If, at any time after the fourth of July and before the actual encounter on the sixteenth between these parties, you heard Greek Crafton say anything about Quinn Harrison, tell where, when and what it was."

Dr. Allen nodded that he understood the question, and that he had something to say about that. "It was the day it happened," he said. "It was early in the morning."

"How long before the affray took place?"

"I can't say positive how long. It was not more than half an hour, I suppose."

"Where was it?"

Dr. Allen had been waiting a long time to tell his story, and finally given the chance it came out in a great burst of words. "I went down to the Plains for my mail and I stopped a little bit at Henry Smith's grocery store and was sitting there in the door and Greek came in and walked back towards where the bucket stood in the middle of the floor and then walked back again and stood in the door and the mail-boy was going along and says he, 'You're early this morning?' Says he, 'Yes, I wanted to start when it was cool and have the cool time for it.' At that he stepped off and was going up towards the post office. Hearing these words, I concluded to go up too and get my mail and I stepped out of the door and Greek was about as far from me as from me to you…"

Hitt estimated "(10 or 12 feet)" and so noted in his transcript. "…when I went out on the walk, but I am before my story. Greek halted when I came out and he heard me and I caught up with him and he says,

'John, I allow to whip Quinn Harrison today.' Says I, 'Greek, what's that all about?' He made no reply to me. Says he, 'I allow...'" Hitt tried to shake the throbbing from his hand, causing him to miss the occasional word. "'...do it this day if I get a chance.' I says, 'You had better let that alone, Greek.' He made no reply, only walked ahead. I went up to Turley's with him and as we went he said he was expecting him in there after his mail."

"Who," Lincoln asked, "Quinn?"

"Yes." A fundamental tenet of the law business is that a smart lawyer gets out of the way once he gets the stone rolling downhill. Lincoln mostly stood aside and let Allen tell his story. "At Turley's," the witness continued, "he made a stand at the door to knock him down as he came in. He was in there a little while and the first thing I saw he pulled off his coat and his hat and laid them down beside the showcase, Mr. Turley's and stood there. But Mr. Harrison did not come in and after a few minutes he went and put his clothes on again."

"Where did he go?"

"The last I saw of him he was pulling off his coat as he walked in where the affray happened. It was not exceeding half an hour before the affray began."

"Your witness, John," Lincoln said.

Dr. Allen shifted in the witness chair, turning his body to face the prosecution's table. Palmer remained seated, appearing to read notes scribbled on a page.

"Do you recall if Mr. Turley was in the store at the time?"

"I don't remember seeing him. It was his store, I can't say whether he was there or not." Like several other previous witnesses, Dr. Allen seemed to have adjusted his attitude. Rather than being free with his words, offering long rambling responses to Lincoln, he was far more guarded in response to Palmer's questions. He would answer the prosecutor's questions honestly, his whole demeanor suggested, but he wasn't going to give anything extra. Whatever Palmer got out of him, he was going to have to ask it out directly.

Palmer knew that. His questions became short and precise: "Who is postmaster?"

"James Ware."

"Who keeps the store, assists Mr. Turley?"

"James Ware is his clerk."

"Were there other persons there at the time Greek Crafton pulled his coat off?"

"Yes, sir, I think Jake Epler was there, and Silas Livergood."

"He walked in and pulled off his coat and put them behind the showcase?"

"Yes, on the west side behind the door. He put his clothes at the south end, some two or three feet from the door I suppose."

"How far was Crafton from the door when Crafton went back and took the place he did?"

"He was standing right at the side of the door."

Palmer asked his questions fast and Hitt wrote as

rapidly as possible, fully aware he was capturing only the words and not the intonation. His hand was throbbing again, but he had willed himself to ignore it.

"In sight of all in the house?"

"Yes."

"Was he leaning against it, or how?"

"He was standing right beside the door." Hitt added that the witness indicated where Crafton was hiding.

"Was he fronting the door?"

"Yes."

"Could a person standing outside have seen him?"

"Not unless they came in right in front of the room, from the north."

"If persons would have come in at the door would they not have seen Crafton?"

Hitt wrote, but he was unable to figure where Palmer was headed. What advantage was it to the prosecution to show that Crafton took a hidden position to attack Harrison?

"Yes, sir, if they had been on the north east side, he stood on the west side."

"What direction did Harrison live from Pleasant Plains?"

"West."

"What direction from Short and Turley's?"

"East."

"You say he stood up beside the door against the wall?"

"Yes, sir, he was close against the door."

The steno man continued writing, unable to see the gain behind these questions. "Anybody by him?"

"No, sir, not that I saw."

"Did he talk to anybody."

"No, sir, I did not hear him."

"Did you?"

"No, sir, I never spoke a word to him."

"You remained there?"

"Yes, I was sitting on the east counter. I don't remember talking to anybody."

"Did you talk to Mr. Epler? After he had started down towards Mr. Short's?"

"Mr. Epler was out on the platform and I walked out."

"Was he on the platform while Greek stood at the door?"

"No, sir, I think he stood in the house."

"How long did he stand there?"

"I don't know."

Mr. Palmer's questioning continued along this line for many more minutes, going over a well-set path time and again without drawing anything of consequence. Greek took off his coat and hat and stood still behind the door. After some two minutes he put them on and left. No one spoke to him. If there was a purpose to these questions, Hitt could not detect it. That was not his job, of course, but as a man who had spent many hours inside courtrooms, he quite often found himself thinking along with a sharp lawyer. He enjoyed watching a skilled attorney navigate through the twists and turns of a case. He wondered if Palmer might be laying groundwork to make a point later that he couldn't quite fathom?

Finally, Mr. Palmer thanked the witness and excused him. Dr. Allen was obviously pleased to be done with it, and scurried through the courtroom and out the side door. He was gone before the next witness was called—John Purvines, a first cousin to gimpy William Purvines. Like witnesses had done since Roman times, John Purvines placed his left hand on the Bible, raised his right hand and promised to "Tell the truth, the whole truth and nothing but the truth, so help me God," and then, as suggested by *Bouvier's Law Dictionary* only three years earlier, he kissed it. While it was probable not one person in that courtroom knew the derivation of the oath, not even the legal scholar Logan, they all held it in great reverence. According to legend, Roman men were compelled to squeeze their testicles while taking the oath, a story explaining why the Latin word *testis* means witness. A man who does this "bears witness" to virility. More likely though, the term witness is from the ancient Greeks, meaning three, a witness being a third person to observe events.

Adding a kiss was far more recent, a gesture of respect.

Abe Lincoln established that John Purvines knew both men and had been at the picnic, and that he had heard Greek Crafton utter threats. "He was talking to Mr. Purvines..."

"Some other Purvines?" Lincoln helped him.

"Yes, a cousin of mine. He came to him and asked him if he knew where Quinn Harrison was. He asked him what he wanted with Quinn. He said he wanted to thrash him, that he had it from Wiley Crafton and he meant to thrash him on that day if he came there."

"Was that the conversation?"

"That was all I heard."

Once again John Palmer took over, asking Purvines to repeat Crafton's words. "He said he intended to whip Quinn Harrison if he came there that day."

Stephen Logan interrupted, asking a question that seemed to come from somewhere over the border. "One question I want to ask somebody. Quinn Harrison has a brother named Peter younger than he is?"

"Yes, sir," Purvines responded. "I suppose he is between seventeen and eighteen years old."

Logan's odd query satisfied, Palmer repeated his original question, what did the witness hear Greek Crafton say? Purvines repeated his answer, "I heard him say he intended to whip Quinn Harrison."

"Repeat his language."

"Says he, 'I'll whip Quinn if he comes here today' and 'I have it from Wiley Crafton to do it.'"

Hitt began to wonder at Palmer's emphasis on the word *whip*. Perhaps, he thought, the prosecutor meant to make a stand on the difference between whip—and kill.

"For what?" Palmer asked.

While this question clearly called for the witness to

state his opinion, the defense let it go, a sign that they did not consider this cross-examination dangerous. "I suppose it's understood what he had," Purvine said. "I suppose he had directions from Wiley to whip him."

"Did he say so?"

"Yes, sir, to Mr. Purvines."

"Was that part of your statement before?"

"Yes, it was."

"Did he say why he was going to whip him?" Hitt noticed more emphasis on that word.

"No, sir. That was all I heard him say."

When Palmer said he was done, Logan said he wanted to ask a few more questions. "Did Quinn come there that day?"

"No."

And then he set the trap. "Do you know the reason why he kept away?"

"Objection!" Palmer caught him long before it might be sprung. He was not quick enough to prevent the jury from hearing the question; although it might be a stretch for them to make the connections. Hitt hid a smile under his palm; the obvious reason Harrison would have stayed away was because he had warning of Crafton's threats—providing him with good reason to protect himself and direct knowledge of the threats.

Judge Rice sustained the objection.

The defense then called Abijah Nottingham, the twenty-three-year-old son of Colonel Jonathon Nottingham. Abijah had grown up on a farm outside

Pleasant Plains and had known both men for his lifetime. Abijah, too, had heard Quinn make threats, he told Lincoln, but he couldn't rightly remember exactly where that was. "I believe my mind is flustered at the present," he apologized, "that I can't tell the month, but it was just prior to the time the difficulty took place between Mr. Crafton and Mr. Harrison; Mr. Crafton did state to me things in reference to what took place. He stated that he had just provocation for assaulting Mr. Harrison and that on the first meeting of Mr. Harrison he should pounce upon him and I think he stated that he should knock him down and stamp him and that he never would suffer such impeachment as had been thrown out on him and his family, or relations and family, I think."

Lincoln urged him to tell his whole remembrance.

"He said that in view of threats that had been made on the part of Mr. Harrison it behooved him to approach him in this manner: that Mr. Harrison had threatened to cut him and to knife him and it behooved him to approach him and not give him any chance to use a knife upon him, and that he, the extent of his threat was to thrash him until he could not go, that his friends had told him never to be implicated in the like that he had been implicated. That, I believe, is the sum and substance."

Lincoln pushed him a bit on his words. "Did I understand you to say that he meant to thrash him so that he could not go on?"

"I think that was the word he used," Nottingham

responded, "or there was something in reference to his laying him up so that he could not go out of his bed or something like that that was the statement he made."

Stephen Logan interrupted to ask, "You say also that he said before that that he intended to knock him down and stamp him?"

"Yes."

Mr. Palmer took the witness. At times Hitt wondered why the other members of the prosecution team did not relieve him from time to time; they all carried a good reputation. But they did not, and truthfully John Palmer seemed none the worse for it. In fact, he seemed to be enjoying it, thoroughly engaged against Lincoln and Logan. He asked a question that had been intriguing everyone. "Did Mr. Crafton say what Mr. Harrison had said about his family?"

"He didn't state what the provocation was."

"Did he say which member of his family he had spoken of?"

"No, sir."

"Did he say himself or his family?"

"He spoke of himself in connection with his family."

"Did he speak of his sister?" Hitt had heard rumors that the marriage of Greek's older brother, William and Eliza Harrison, had gone sour. There was talk of hitting. But the details had been sketchy, at best, and it was unreliable information.

Nottingham was not the one to clear it up. "No, sir, he did not. He only mentioned himself and said he would never be impeached in the light or implicated in the light as all of his family had been. He never would be implicated in the light that Harrison had implicated him and his family. That was the reason."

"You said that he said that Harrison was armed?"

"Yes."

"And that he should approach him in such a way that he could not use his knife, of this he would be certain to do."

But Nottingham never said right out that he had told Harrison of this conversation. It was becoming apparent that Lincoln and Logan were struggling to prove that Harrison had learned of Crafton's threats. Had the law been more tolerant, they simply could have put Quinn Harrison on the stand and have him tell his story. But as that was not permitted, they did their best to show that the entire community was aware of those dire remarks. It is quite possible that Palmer, too, regretted the fact that Harrison could not testify. As history has shown, defendants take the witness stand at considerable risk, as the cross-examination by a deft prosecutor, Mr. Palmer for instance, might shred their story, leaving the defense in tatters.

Next called was James S. Zane, another of the young farm boys of Sangamon County who had grown up beside Crafton and Harrison. Zane explained that he had heard Greek threaten Quinn Harrison several times. "The first that I heard him speak about it was at Clary's Grove, on the fourth of July. Another time I heard him speak about it in company with…"

Changing papers, Hitt missed the proper name, but resolved to fill it in later. "…coming over from his father's to Pleasant Plains." On the fourth, the witness continued, "He said he had difficulty with Harrison that morning and that he intended to thrash him and so on."

Lincoln asked him to "Fill up the 'so on.'"

"He said he 'intended to maul him,' was the word he used at that time, I think, as near as I can remember."

"If that is all you can remember of that time, tell us when the other time was and what he said."

Zane looked at his friends in the gallery and smiled broadly. It was a smile that said he was enjoying his moment, and as far as he believed, he was doing pretty good with it. "It was on Tuesday before the affray. It was on the road betwixt Mr. Crafton's residence. At that time he said he was going to stamp him."

Lincoln had a knack for picking up a witness's steam and removing obstacles. His supportive attitude made witnesses want to tell him their story. "Repeat his language as near as you can, the whole of it?"

"I was telling him, we were in conversation about it, and he allowed he would knock him down and stamp him was the words he used as nigh as I can say."

Lincoln needed his support, then asked, "Did he say whether he expected Quinn to be armed or not?"

"Yes. He said, 'Damn him.'"

Lincoln asked several additional questions, but this was the essence of Zane's direct testimony.

Palmer could not match Lincoln in his ability to relate to his witnesses, so instead he brought with him a pure officiousness that at times appeared to intimidate witnesses. He began his cross-examination by asking, "Did he state in either of these conversations why he meant to maul him?"

"He said Mr. Harrison insulted him."

"How?"

"In talking regarding his brother. Mr. Harrison reproving his brother for keeping company with him, as I understood it, I think."

"Do you recollect what he said?"

"No, sir."

Palmer offered help. "Do you not recollect he spoke of the insult to his family?"

Zane admitted, "I might have heard something about it. He said he was insulted that Mr. Harrison had called him a 'D—d son of a b—h.'"

"Didn't he say that Harrison had called him a d—d spreckled-face son of a b—h?"

"I don't think I ever heard *that* expression."

"Did he tell you about Harrison threatening to cut him in the morning?"

"Yes. He said Harrison threatened to kill him if he jumped on him."

As Hitt wrote, he wondered if Palmer might not have made an error pursuing this line of questioning. Clearly this implied some degree of communication existed between the two men. Otherwise why would Harrison find it necessary to protect himself "if (Crafton) jumped on him?" He must have had some reason for anticipating this attack?

"Didn't he say that in an affair between him and Harrison they would have come together if he hadn't been for Henry?" Hitt thought it odd that Henry was not further identified, and wondered if he had misunderstood the word. He suspected Zane might be referring to Fred Henry, but it disturbed his sense of order that it was not pinned down.

Zane didn't object to it though, responding, "I don't know as he stated it at that time. I never had a talk with him about that. I don't know what he said they would have done if they had come together. I think the words he used was that Harrison had called him a 'd—d son of a b—h.'"

After only a few more questions, which yielded nothing more of value, Lincoln faced the bench and announced, "Our evidence is all presented, your Honor. The defense rests."

Once again, the courtroom came to life, as if a withheld breath suddenly had been released. Judge Rice gave them a minute, then banged his gavel. "We're not done yet," he told them, the vibrations of the gavel jolting through his back. "Mr. Palmer still has a couple of people he wants us to hear from. So quiet on down please."

Palmer thanked the judge, then began his rebuttal case, calling Jacob Epler to the stand. Although in his fifties, Jake Epler was still a vibrant fellow, well-liked and respected by his neighbors. Epler had moved to Sangamon County in 1848, opening a general store a year later where the stagecoach out of Springfield to Beardstown crossed Richland Creek. People had been settling in the area since 1819, when Mr. Spillars built a horsepowered gristmill. A few years after Epler opened his place, John Adams came in with a blacksmith shop and the town just grew up around those two stores. Epler gave it the name Pleasant Plains, although it wasn't incorporated for several more years, taking the name from a small Methodist church that, ironically, Peter Cartwright had founded and built in 1838, just a few miles north of Epler's store. Since then, it had become a prosperous community.

The townsfolk knew and liked Jake Epler; through the years he'd extended credit to many of them, helped them ride through the tough times. He stood by his word, which made it every bit as valuable as that of Peter Cartwright. For some people, maybe even more: a lot of them entrusted their eternal soul to Pastor Cartwright, Jake Epler had their money in the here and now. Epler told John Palmer that he knew both boys, and said he had spoken with Greek Crafton on the day of the incident, but only after the cutting was done.

"Do you know when Mr. Peter Cartwright was there?"

"Yes, sir. I was not there when Mr. Cartwright was

there," he said. Then after some thinking corrected himself, "Yes, I was there too, where he was there."

Palmer stopped here to give Judge Rice fair warning. Mr. Epler was going to testify that Greek Crafton had some other things to say after the Reverend Cartwright departed in regard to the cause of his death, and these words were not so forgiving of Harrison.

Lincoln objected vehemently. They were back and forth with it, each man making a good case for his position. As in many legal matters, Hitt knew, both sides in this argument had merit, and the judge would have to shave it closely however he decided. Judge Rice took the same position he had held earlier: let's hear the testimony without the jury and then figure it out. For a second time the jury was led out of the room. Several of them were smiling, Hitt noticed, perhaps remembering they were given food and drink while they were holed up in that room. When they were gone, Palmer asked Epler to "State if you heard Mr. Crafton say anything about the connection of Mr. Harrison with his death, after the time Mr. Cartwright was there and held his conversation with him."

Mr. Epler had a soft phlegmatic voice, causing several people in the gallery to cup both ears with their hands so they might hear him better. "It was after the conversation with Mr. Cartwright," he said, referring to the first of several conversations Cartwright would have with Greek Crafton. "But I would have to bring in about Mr. Short's coming in to make it plain."

"Yes, go on."

"Well, Mr. Short came in. That is Mr. B. J. Short that you have had before you, I presume. As he was coming in, Greek from where he lay, could see him coming through the gate and he just named him, 'If it hadn't been for that man, Mr. Short, he wouldn't have been there.' He just said that and that was about what he said in regard to that..." Mr. Epler hesitated, trying to remember Crafton's exact words. In deference to his position the court remained silent until he was ready to go on. "I want to have it right," he said. "You've got me bothered. Well, I don't know that there was anything more said in regard to Harrison."

"Is that all you remember?" Palmer asked sympathetically.

This was a very important point. According to Epler, Greek believed that Short's interference had effectively caused the tragedy. Epler was struggling. "No, sir," he finally said, "not all I can state what he said. That was all he said, I believe, to Mr Short. He was asking after that whether Harrison had yet been taken by authority." Mr. Epler recalled that being important to Crafton, had Harrison been arrested? He asked it several times.

Both Lincoln and Logan stood at almost the same instant, and then looked at each other, somewhat surprised and pleased, and smiled. Before those smiles disappeared the judge motioned for them to sit, allowing Epler to finish. After all, this was being heard outside the presence of the jury. "He said it was true that Harrison was taken into custody. That's all. That's the sum and substance."

Mr. Logan now made his argument, stating loudly that the jury should not hear this because it was not at all rebuttal testimony; far as he could see it did not contradict Pastor Cartwright's recounting of the facts. Therefore, it did not belong at this portion of the trial. Mr. Epler should have testified when the prosecution made its case, Logan said, and permitting him to be heard at this time might lead the jury to false conclusions. The judge reserved his ruling for now and asked the parties to move on.

Palmer announced that the prosecution would like to recall Dr. J. L. Million, who would testify that during all the hours he had spent at Quinn's bedside he had never heard him, in any way, forgive Peachy or put blame on himself. It made sense, he said, to hear both of these witnesses before the judge made up his mind. The defense agreed that made sense, and Dr. Million was recalled.

Dr. Million sat straight up, almost rigid, in the chair, halfway back in the seat, his elbows sitting on the armrests with his arms lying flat their whole length so his hands could grasp the ball handles. Palmer went right at him, telling him to "State whether you had any conversation with Greek Crafton after Mr. Cartwright had the interview, about the connection of Mr. Harrison with his death?"

"Not after," Dr. Million corrected him. "Before Mr. Cartwright had his interview."

"Did you hear him say anything about Quinn Harrison after that conversation?"

"No, sir, I don't recollect hearing him say anything after Mr. Cartwright was there."

"Did you hear him say anything about it before Peter Cartwright was there?"

"Yes."

"What was his condition at the time?"

"I would say he was rational."

When he was done the defense objected to the testimony of both men. After pondering this for a brief time, Judge Rice pointed out that Greek Crafton's statements to Mr. Epler and Dr. Million, as provocative as they might seem to be, were made prior to the crucial remarks to Pastor Cartwright, and before he was aware his wounds were to be fatal. As such, according to the law, they did not qualify as appropriate rebuttal nor a dying declaration and therefore he had no choice but to sustain the defense objection.

Upon hearing that ruling a disappointed Palmer announced that he was withdrawing the testimony of both witnesses. And with that, he said, he had no more evidence. He looked up at Judge Rice, spread his palms in a gesture indicating he'd finished, and said, "The prosecution rests, your Honor." His case was done.

Judge Rice informed both sides that the court would resume the following morning at 9:00 a.m. when summations would begin. He brought the jury back into the courtroom and reminded them that they weren't to talk about the case with anyone, absolutely anyone, "Even if she threatened she wasn't going to fix your dinner," which drew a hearty laugh, and then he wished everyone a pleasant evening and banged the proceedings closed.

It was early evening by the time Hitt had packed his equipment and walked out into the twilight, the heat and humidity still hanging listlessly. The telegraph was working once again, he had been informed, and he wanted to stop by the office to collect any messages. Then he would return to the Globe to work on his transcriptions. He wasn't Lincoln or Logan, Palmer and all the rest of them. All men he admired for their intelligence and fortitude. He accepted that. No lives depended on his work. But in an important way, he believed, he was making a valuable contribution to the system: he was bringing the words of justice being done to the American people.

CHAPTER FOURTEEN

Trial lawyers are natural storytellers. As Hitt had learned very quickly upon entering his profession, in addition to a thorough understanding of the statutes, good lawyering required the ability to spin a tale, create a mood or magnify an emotion. Sometimes a lawyer had to speak fast and smart to cover some bad ground, while at other times he had to go slow and smooth so not a word was missed by a juror.

Lawyers especially liked to talk about their colleagues. In a profession of constantly shifting alliances, they needed to know the strengths and weaknesses of each man they might someday work with, or go up against. When Abe Lincoln made his name by crisscrossing the state of Illinois debating Senator Douglas, the Chicago lawyers all told their stories about their dealings with him, unabashedly hooking themselves to his rising star. At different times they would corner Hitt, who had come to be associated with Lincoln during those debates, to swap a few tales. It seemed like every lawyer in Illinois had suddenly become an expert on Lincoln. In fact, after two decades' worth of cases, many of them had been involved with him, often more than one time, and they all held a pretty strong opinion. Admittedly that opinion varied depending on their political position. But to Hitt, the consensus seemed to come down to this: there was no better lawyer in the state when it came to standing in front of a jury and summing up his case. Closing arguments were the star turn for a lawyer, an opportunity to speak at great length, with the whole of the courtroom in his hands. Until this time a case was a battle of facts, focusing on cross-examination. But here all the skills of a lawyer had to be brought into play; the ability to interpret facts in a way most favorable for your case, a time to combine intellect and passion, character and justice. It was also the final chance to speak directly to jurors. Lincoln had learned how to talk to a jury on the circuit, where juries made their decisions as much on emotion as the law. He had figured out how to reach out and grab hold of those emotions and turn them on his client's behalf.

And as Hitt got ready to hear the closing arguments

in this case, he remembered some of those comments he had heard. According to Leonard Swett, a man himself admired as one of the most brilliant members of the bar, "If he had any superiors before a jury—and the more intelligent the jury the more it pleases him—I never heard of him. I often heard Tom Corwin, Sargent Prentiss, Rufus Choate and Humphrey Marshall, but Lincoln at his best was more sincere and impressive than all of them, and what he cannot accomplish with a jury, no living man need try."

Even his adversary Stephen Douglas was an admirer, saying publicly, "As a pleader before a jury he seems in congenial relation at once; and before any jury I have ever seen him address, there was little use for any lawyer to oppose him except in matters of fact, and in those he always conceded the truth in the cases which he tried; but as an advocate he had no equal before a jury, and if he has ever met one, I have never heard of it."

Lincoln's good friend Judge David Davis said, "He seized the strong points of a cause and presented them with clearness and great compactness. His mind was logical and direct, and he did not indulge in extraneous discussion." Judge Davis loved to tell people about Lincoln's closing argument in the Wright case, a case they had worked together. They represented the widow of a Revolutionary War soldier who had been granted a $400 pension for her husband's service to the nation. When an unscrupulous agent named Wright withheld $200 of that blood-earned money,

Lincoln and Logan sued on her behalf. The day before his summation, Davis said with a smile, "Abe told me to make sure I was in the courtroom, because, 'I am going to skin Wright, and get that money back.'"

Lincoln had done his homework for the summation, Davis recalled. More important, he used his natural gift for storytelling to yank on the jury's every emotion. He talked of the injustices that led to the war, then drew a detailed portrait of the hardships those patriots endured in the winter of Valley Forge. Barefoot and with bleeding feet, these heroic men crept over the ice to put their lives in jeopardy for an idea. And it was their suffering that gave us this nation.

That part done, he set after the agent. Davis would describe Lincoln as holding a handkerchief in his right hand, then hold out his own right hand, and when he said Lincoln threw that handkerchief aside he would throw his own imaginary handkerchief. And then, he recalled, Lincoln began an attack on the agent that was "hurtful in denunciation and merciless in castigation." With no boundaries set by the judge, he tore into this despicable man as if he had been fighting to prevent the birth of this nation.

Hitt had been told repeatedly that no lawyer closed his case better than Lincoln, and Davis certainly supported that contention. Lincoln was standing in front of the jury, as Davis told it, and describing for them the moment that this young man kissed his young bride and babe in the cradle and set off to war with nothing more than the clothes he was wearing, his

ABRAHAM LINCOLN PRESIDENTIAL LIBRARY AND MUSEUM

This photograph of Lincoln was taken in Chicago in early October 1859, a month after the conclusion of the trial. While many people in Springfield were unhappy with the verdict, there was little animosity directed at Lincoln.

musket and a horn of powder. "Time rolls by," Lincoln said, "the heroes of '76 have passed away and are encamped on the other shore. The soldier has gone to rest, and now crippled, blinded and broken, his widow comes to you and me, gentlemen of the jury, to right her wrongs. She was not always thus. She was once a beautiful young woman. Her step was as elastic, her face as fair, and her voice as sweet as any that rang in the mountains of old Virginia. But now she is poor and defenseless. Out here on the prairies of Illinois, many hundreds of miles away from the scenes of her childhood, she appeals to us, who enjoy the privileges achieved for us by the patriots of the Revolution, for our sympathetic and manly protection. All I ask is, shall we befriend her?"

Davis savored the memory. Half the jury was crying, the other half was furious. They awarded the woman every cent she was due. More than that, Davis added, Lincoln paid the widow's way home and hotel bill while she stayed in Springfield and refused to accept any fee for his services.

In addition to his use of colorful imagery, Lincoln at times would use props in his closing argument to make his point. In the medical malpractice suit, *Fleming v. Rogers & Crothers*, he represented a doctor accused of causing permanent damage by improperly setting carpenter Samuel G. Fleming's broken legs. In his summation Lincoln used chicken bones to demonstrate the fact that leg bones from a young chicken were supple and would flex and bend, while the leg bones from an older chicken were brittle and would snap far more easily. This display was said to influence the jurors, who were naturally disposed toward their injured neighbor rather than wealthy physicians. In the end, the jury was unable to reach a decision, but after considerable legal maneuvering, a fair settlement was reached making a retrial unnecessary.

While Lincoln's dramatic presentation of an 1857 almanac in the Duff Armstrong case, revealing that the moon had set before the eyewitness claimed its light had enabled him to see the murder, was fast becoming legend, the state's prosecuting attorney, J. Henry Shaw told people it was his summation that won the jury. "His rendition of his childhood struggles, which had been made tolerable by the kindness of the Armstrongs, brought tears to Lincoln's eyes. The sight of his tall, quivering frame and the particulars of the story he so pathetically told, moved the jury to tears also, and they forgot the guilt of the defendant in their admiration of his advocate. It was the most touching scene I ever witnessed."

As Hitt knew from his own experience, Lincoln's summation skills were not limited to emotional appeals. He was equally adroit at putting highly technical information into sensible language that plain folks could understand, and doing so in a way most beneficial to his client. Hitt had seen that firsthand in the *Effie Afton* case, where the owner of a steamboat that had crashed into bridge pilings and been lost sued the builders of a bridge, claiming it was a navigational hazard.

Hitt had met Lincoln during that trial and had been greatly taken with Lincoln's approach. Lincoln had done intensive studies of river currents and navigational skills and made a strong and logical argument that a better man at the wheel of the boat would have handled the situation with ease, but it was his summation that left Hitt awestruck. He began his presentation with a logical, cogent argument that summed up the details of the event and showed that the boat pilot was at fault, but then took it to the next level by grabbing hold of the jurors' dreams, making them understand that the very future of this country might well lay in their hands. They were not deciding who was at fault in a simple river accident, they were being called upon to determine if tradition was to halt progress.

The attorneys representing the boat owners were far less prophetic, claiming that a poorly designed and built bridge was the cause of the accident. A simple case of negligence. Lincoln responded by telling the jury, "A suspension bridge cannot be built so high but that the chimneys of the boats will grow up till they cannot pass. The steamboat men will take pains to make them grow." It was the obligation of the plaintiff to prove that the bridge was a "material obstruction and that they have managed their boat with reasonable care and skill…"

Later he returned to that, asking, "What is reasonable skill and care? This is a thing of which the jury are to judge. I differ from the other side when it says that they are bound to exercise no more care than was taken before the building of the bridge… It is unreasonable for (the pilot) to dash on heedless of this structure, which has legally been put there. The *Afton* came there on the fifth and lay at Rock Island till the next morning. When a boat lies up the pilot has a holiday, and would not any of these jurors have then gone around the bridge and gotten acquainted with the place?"

If there was negligence, he insisted, it lay at the foot of the pilot. After pointing out that another boat had passed beneath the bridge in the same current without difficulty, he said the *Effie Afton* got "so far wrong that she never got it right. Is the defense to blame for that?"

It was men of vision who built this country, he said, and "this bridge must be treated with respect in this court, and is not to be kicked around with contempt... The proper mode for all parties in this affair is to live and let live..."

Hitt had been completely absorbed by Lincoln's summation, and had he been seated in the jury box, he didn't doubt his vote. The case was dismissed when jurors were unable to reach a decision and was never brought to trial a second time, a victory for the wheel and rail industries.

So everyone anticipated that Lincoln's closing argument in the Harrison case was going to be a humdinger. While everyone knew John Palmer was no slouch, his oratory skills were no match for Lincoln's. A great closing argument was top-notch entertainment, liable to be talked about and dissected for days afterward, and this was Abe Lincoln's forte.

While few people thought about it, the concept of a closing argument, a summation, was a relatively recent addition to criminal trials. In England, it was only two decades earlier, in 1836, that the right of the defense to make a closing argument to the jury had first been recognized. Prior to that judges had done the summing up, and often by their words let the jurors know what verdict they desired. In fact, even the appearance of lawyers in a legal dispute was barely a century old, and it had come about mostly in civil trials where merchants and shopkeepers needed people familiar with the laws to represent them in business disputes.

As closing arguments evolved into an important element of a trial, rules developed restraining what might be said. As long as lawyers stuck roughly to the evidence, there wasn't much they couldn't say. Most summations consisted of a presentation of the facts, a rebuttal of the opponent's case and an emotional appeal to reason. Lawyers were not allowed to bring in facts not presented during the trial, misstate the testimony of witnesses or put words in a man's mouth. Lawyers were allowed to discuss the law itself, but had to be entirely accurate about it, and they were not allowed to suggest to jurors that they follow their heart rather than their conscience; on occasion lawyers had tried to quote the Ten Commandments in summations, but they had often been stopped from doing so, as judges had pointed out people were being tried in this world, whatever plans the Lord had for them was His business. Attorneys were allowed to personally display as much emotion as they might be able to conjure. Lincoln had been known to shed tears for example, but they couldn't plead for sympathy; they weren't supposed to point out the crying parents of the accused, or talk about how the defendant was driven to crime by hunger. Nor were they permitted to draw attention to a defendant's ethnicity, race, nationality or religion. Other than those prohibitions though, a lawyer was limited in his talk only by his own imagination and powers of persuasion. And at that time, judges were a bit more lenient about enforcing all of those rules anyway.

The final day of the trial promised to be a memorable event for those people lucky enough to squeeze into the courtroom. When Hitt stopped at the telegraph office that morning to send off his work, the key man told him he hadn't seen so much excitement in the city since the acclaimed actor Charles Walter Couldock and his traveling stock company had done *Richelieu* in Metropolitan Hall four years earlier. Although, he added, people were also well stirred when the noted dancer, actress and mistress of the King of Bavaria, Lola Montez, had come to Springfield to give a lecture on fashion at Cook's Hall. He emphasized the word *dancer* by raising his eyebrows, to make certain Mr. Hitt caught his meaning.

Of all the days of the heat wave, this was by far the worst. By comparison the previous days were just a little warming up. But that did not seem to deter any interested spectators from turning out. The line stretching from the courthouse steps was already far longer than the spaces available when Hitt got there. The courtroom was stifling. Several volunteers were busy fanning the room, trying to get the air circulating before they let the people in.

While Hitt was not engaged to transcribe the closing arguments, he would not have missed being there for the conclusion. Judge Rice laid down the rules as if he was officiating a bare-knuckle fight. The prosecution would have first say. He wasn't going to put any limit on the number of lawyers who might speak or limit their time, but he did ask that both sides be considerate of the well-being of the jurors and the spectators, "and your Honor."

Hitt observed that the lawyers had dressed for the event, each of them wearing a coat and stock tie; several of them even had vests. Clothes made an impression on juries, and these men were wearing serious clothes.

Norman Broadwell spoke first for the prosecution. He stood behind their table, referring often to his notes. His part in the summing up was to rehash the facts of the case and explain how the law applied to them. As he began his carefully prepared presentation, Hitt wondered how Lincoln looked upon his former student, and guessed he took substantial pride in Broadwell's growing success. Lincoln had sent quite a number of good young lawyers into the legal world, although sadly Greek Crafton was not to be among them.

The fact that Broadwell referred to his notes when

speaking obviously was not something he had learned while training with Lincoln and Herndon. Lincoln was renowned for his ability to speak spontaneously, and had once said quite famously, "Extemporaneous speaking should be practiced and cultivated; it is the lawyer's avenue to the public."

Broadwell spoke for almost an hour; he reminded the jury what each witness had said: Greek Crafton had died slowly from knife wounds inflicted by the accused, Quinn Harrison. The dispute had been kindled at the town picnic on the Fourth of July, when Harrison was heard to warn his brother to stay away from the Craftons. Over the next days the two young men told friends they were preparing for a fight although, and Broadwell emphasized this point, the defense had presented not a shred of evidence that Harrison had direct knowledge of Crafton's remarks. Conversely, Crafton had been warned that Harrison was getting a weapon, a gun or a knife, yet still came to the fight completely unarmed. It was clear he intended this to be a fair fight, no different than the countless fights between spirited young men that take place every single day. It was only the handiwork of the accused, Quinn Harrison—and with that he turned dramatically and pointed at Harrison—that brought everyone into this sweltering courtroom. Hitt glanced at Lincoln, who was sitting perfectly still, his head tilted back, his face showing no movement, no response, no emotion; his eyes so completely fixed on a spot on the ceiling that Hitt had no choice but to follow his gaze upward to see what had caught his attention. There was nothing there, nothing at all, and the scribe realized this was Lincoln in full concentration, seemingly oblivious to anything happening around him other than Broadwell's presentation. And his mind was locked into that.

Broadwell then explained that the law might not always be in accord with each person's feelings about right and wrong, but because those beliefs varied so much—even among the wise men of this jury—the law provided a measuring stick that everyone could follow. And that law offered no protection to Harrison; it didn't say anywhere in the law you could stick someone who was hitting you. What it did say is that you can use lethal force only when absolutely necessary, like when your life was in jeopardy, and throughout this entire trial there was not even a suggestion that Quinn Harrison's own life was threatened, he argued. Even that testimony by the Reverend Cartwright in defense of his grandson did not mention any claim from Greek Crafton that he intended to cause serious harm to his adversary. For all those reasons, Broadwell said, his voice rising to a pitch, the members of the jury had no choice but to find the defendant guilty. He sat down. The courtroom was respectfully silent.

Broadwell had set a high standard; his argument was logical and factual. The defense was going to have to go some to knock it down if Harrison was to be saved. They elected Shelby Cullom to make their first statement. It was Cullom who had hidden Harrison under floorboards after the fight, and now he was to begin the summary argument for the defense. Several of the jurors had watched the thirty-year-old Cullom grow into manhood, others of about the same age had worked with him; in small cities like Springfield it would have been almost impossible to seat a jury without some connections to the participants. Cullom was the well-liked city attorney, and respected throughout the city. The kind of young man that led people to point him out as a man with a good future, and his career had been pushed along by men like Lincoln and John Palmer. So when he stood for the defense, the jurors and the gallery of spectators listened carefully. Cullom stood near the jury and began by laying out the defense version of the facts. There was bad blood between the Crafton and Harrison families. Words were spoken. Greek Crafton was pushed by his father to...

Cullom faltered. He coughed thinly and asked the indulgence of the court. He sat and sipped some cooled water. When he had rested he stood up and began again. Words were spoken between Crafton and Harrison. Greek Crafton was pushed by his father to defend the family honor. The jury had heard for themselves how many people Crafton had told he was going to stamp and maul the much smaller Harrison. He...he...

The attorney sat down again. His color was drained. He clamped his palm over his mouth to prevent an accident. The heat maybe combined with some nerves, had consumed him. Mr. Logan stood and said he would take charge. Mr. Cullom weakly thanked him and then with some help left the courtroom. Later, he would write about this moment: "The courthouse was crowded with people and it was hot and smoldering. I was so overcome with the heat and the great responsibility I thought was resting upon me as a young lawyer I broke down in the midst of my speech and practically had to be carried out of the courtroom."

Stephen Logan, as always, was completely prepared and in a thorough and utterly professional presentation laid out in simple terms why the jury must find Quinn Harrison not guilty of murder.

Like Broadwell, Logan reviewed the facts that had been heard in the courtroom, but not surprisingly his interpretation of those facts differed in almost every way from the prosecution. The slight Harrison had almost got into a fight with Crafton at the town picnic, but the young men had been separated before harm was done. As the jury could plainly see just by looking at him, Harrison was small and slender, while witnesses described Crafton as a much larger and stronger man; in addition, Harrison also had been ill for a considerable time, an illness that had sapped his strength. He did not set out the morning of the sixteenth to confront Mr. Crafton—he was not looking for a fight. He did not see Greek Crafton, who had lain in wait to surprise him at Mr. Turley's store and followed him to Short's. It was there, as he was sitting peacefully reading a newspaper, that he was attacked and grabbed from behind. He did not try to fight back, instead he held on to the rail so tightly that the entire counter was yanked into the middle of the floor. He even called out for help: numerous witnesses heard him scream "have I no friends here?" To no avail. It was only after he had been pulled free, and it appeared that John Crafton might join the fray, that in fear for his life he had struck back with the only weapon he had available to him.

Greek Crafton's intentions that day were well-known; he had been telling everyone who would listen to him what he was going to do to Quinn Harrison. He was going to stamp him! He was going to maul him! He had been given leave by his father to defend the family honor. He had promised he was going to hurt Quinn Harrison, but only he knew what that meant. There could be no doubt as to his frame of mind when he attacked Quinn Harrison that morning. The anger that welled inside him was bursting to get out. He had put his own reputation and that of his family on the line, anything less than a beating that would satisfy that thirst for revenge would not be sufficient. At this point Logan began referring directly to the Reverend Cartwright's testimony. He had written down a few of the phrases he had used and quoted them. In his dying declaration Crafton admitted he had started this and had come to regret it. But in his last moments he found extraordinary courage, and accepted responsibility for his actions. Logan gave credit to Greek for coming clean in what must have been a difficult confession. After Logan's dramatic re-creation of Cartwright's testimony, Judge Rice decided he needed a break from the heat. He interrupted Logan and asked him to pick up again after everyone had an opportunity to cool down a little. Having already spoken for more than an hour, he readily agreed.

Hitt watched as Lincoln and Logan escorted Harrison to their waiting room. It occurred to him at that moment that he had never heard the sound of Harrison's voice. That struck him as odd; there was the most important figure in the entire trial, the very reason for the whole assembly, and he had not spoken one word. He barely had spoken to his attorneys. Different countries follow different rules for trials; in some European countries the accused was given an opportunity to speak directly to the jury at the end of a trial, but a quirk in American jurisprudence at that time kept arguably the most significant voice silent. Hitt wondered what he would say if given the opportunity. Would he claim he acted out of terror? Would he apologize? Would he say it was a terrible accident? How would the jury react? Would those twelve men believe him? Or would they dismiss his words as self-serving and untrue?

Had Harrison been permitted to testify, there is no doubt the trial would have proceeded differently. His story would have been the spine that the defense would have tried to support and the prosecution would have attacked. But the law left it to others to make the case, while the man whose life was at stake sat there mostly unnoticed.

Like so many other structures in the young nation, the legal system was slowly evolving. A ragged collection of state and territorial laws, which had been set in place to serve the specific needs of that region, was being transformed into an orderly and predictable national system. There was still a long way to go, one state's assault was another state's self-defense, but progress was being made. In some areas logic and reason had already replaced tradition. Hitt's own place in the courtroom, writing down the exact words so anyone at any time—even over 150 years later—could know what was said here, was itself a significant advance. The steno man never flattered himself that he was writing history; he was just doing his job, just recording the words. But having a true record, being able to look back on his transcript and know this is what was said in this courtroom on this day, added immeasurably to the confidence Americans could have in their courts. It was a revolutionary development, shedding light on what for too long in too many places had been a process done mostly in the dark.

Except, of course, for the opening and closing arguments. Those were attempts by the prosecution and defense to persuade jurors to come their way, but they weren't evidence and so, for now, it didn't seem necessary to record them verbatim.

After the break Stephen Logan, refreshed, resumed his final argument. Even Shelby Cullom was back in the courtroom, sitting behind Lincoln and Logan. It appeared to Hitt that the room was even more crowded, if possible. Logan resumed by once again focusing on Cartwright's testimony. Although he was barred from saying outright that the Reverend Cartwright was not capable of lying, even to save his kin, he cleverly maneuvered around those restrictions to make the point: Greek Crafton had accepted responsibility for the fracas. If he had forgiven the defendant, the jury must do no less.

As had Broadwell, Logan went through the applicable statutes to show that Harrison had done nothing illegal by arming himself, although he was careful to avoid saying specifically that his client had knowledge of the threats. No evidence had been presented to prove that, so he couldn't bring it in out of thin air. He reviewed those clauses covering self-defense. The right of a man to protect himself is a fundamental principle on which this nation had been founded. You have a right to be secure. Was this slender young man simply supposed to stand there until he was beaten senseless by a bigger, stronger attacker? By two attackers? He did everything in his power to avoid the fight. He held on to that counter as if holding on to his life. Only after he was yanked free, while one arm was being held tight, when he was in the midst of a melee, Greek Crafton holding him, Short holding on, John Crafton entering the fight, did he reach his last resort and go for his knife.

It was, the *Register* would report, a grand performance. "Judge Logan...continued for two hours with great ability, discussing the facts with the most impressive earnestness and eloquence." He finished with a plea to the jurors that they put themselves in the position of young Quinn Harrison, and ask themselves if they had been caught in that situation, what would they have done? Logan had confidence that the men of the jury, several of whom, most likely, at some time in their lives had found themselves in a similar predicament, would have fought back to their greatest capability. What happened was truly a tragedy, no one could dispute that; a fine young man had lost his life in a completely unnecessary fight, but the law spoke quite clearly on this matter. For all those reasons, Logan said, the members of the jury had no choice but to find the defendant innocent.

He was sweating profusely when he sat down. He wiped his face with a dampened cloth as Lincoln and then Cullom congratulated him. Judge Rice gave the courtroom several minutes to quiet down, as spectators made their opinions known to their friends. Logan had been a match for Broadwell; both men had made salient points and done so with conviction. It was a close contest, that was for sure.

Finally Judge Rice looked at the defense table, nodded and said, "Mr. Lincoln."

Abraham Lincoln spent several seconds straightening the papers on the table in front of him, took a moment to fix his stock, then stood ramrod straight to the full of his great height and faced the jury. And smiled.

CHAPTER FIFTEEN

Abe Lincoln had the knack for making his complete preparation look casual, from the choice of his words to the cut of his cloth. He dressed to satisfy the jury. To his closing argument in a high-profile libel case argued in Vermillion County, where he was not well-known, he had worn a fine broadcloth suit, a silk choker and thoroughly polished boots. But here in Springfield the people knew him well and would not be impressed by any highfalutin silk ties. So he wore his almost-threadbare black frock coat, a vest, an old-fashioned stock tie and handmade suspenders, and shapeless brown trousers. His boots were dull from use, although they had been soaped clean. He was a Sangamon man.

He walked to the jury box and took in all twelve of them with a glance. He wished them "afternoon," and said hello to those five or six men he knew by first name. He was following the first rule of good criminal lawyering, Hitt realized; he was building a relationship with the jury. He was just Abe, their neighbor, the man who shared their values and their lives, standing here hopin' they could solve this sticky problem together. Years earlier, Hitt knew, Lincoln had given practical advice about talking to a jury to a young man he was mentoring, it might have even been Broadwell, and that advice had been passed along and passed along until he had heard it. "Talk to the jury as though your client's fate depends on every word you utter. Forget that you have any one to fall back upon, and you will do justice to yourself and your client."

He spoke to the jury in a soft voice, often dropping the final *g* of a word, and yet it did not come across to Hitt as an affectation or a ploy; rather a tone the *State Journal* would describe as "earnest, natural and energetic." And it was natural to him; like the clothes he wore; he spoke the language of the jurors. In his advice to his partner, Herndon, he had cautioned him, "Billy, don't shoot too high; aim lower and the common people will understand you. They are the ones you want to reach. The educated and refined people will understand you, anyway. If you aim too high, your idea will go over the head of the masses and hit only those who need no hitting."

To make a connection to those common people, Lincoln quite often began his closing arguments with a story, most of the time told at his own expense. He begged the indulgence of this jury, for example, because he had a lot to say and sometimes he could be a bit forgetful. In that he was like the old Englishman "who was so absentminded that when he went to bed he put his clothes carefully into the bed and threw himself over the back of his chair."

The easy laughter from the jury was an indication that he was forming a bond with them. While Hitt wrote none of it down, it is likely based on newspaper reports from the time and everything we know of his career, that Lincoln started by reminding jurors what an awesome obligation lay before them. Just as he had done in so many other final remarks. Unlike many lawyers, he did not read from the applicable statutes or quote authorities; instead with a sweep of his hand he took in the bench and the prosecution, and said, "These gentlemen will allow, or Judge Rice if need be, will explain the laws that are to be applied here." It was his intention to remind them that they were holding a life in their hands, and so they needed to be sharp, they needed to listen and understand every word, because one word might make all the difference in their deliberations. Quinn Harrison lay accused of committing the most heinous of all crimes, murder. Murder is a legal term. It is strictly defined by the statutes. But it comes down to more than a legal definition: to find Quinn Harrison guilty of murder, as opposed to any lesser crime, every single juror had to be convinced that Quinn intended to grievously harm Greek Crafton rather than merely defend himself. And Lincoln argued, there was nothing in the prosecution's case that supported that contention.

At the very beginning of his presentation, Lincoln stood almost stoic, his hands locked behind him, his left hand held in his right hand. But as he warmed he became more animated; at first he did not use his hands to emphasize his point but then he brought them in front of him, interlocking his fingers, and gradually began using them in common gestures. When making a point, he suddenly would shoot out his long forefinger, an action so completely unexpected it was impossible not to follow its line to the intended target. At those moments, Herndon said, "Every organ of his body was in motion and acted with ease, elegance, and grace."

Once in motion, he remained in motion, moving around the courtroom but always, always, returning to a place in front of the jury. The first part of his argument was a lively discussion of the evidence and the testimony the jurors had heard, and piece by piece he picked it apart, highlighting the inconsistencies, pointing out the gaps—although he was careful never to question the truthfulness of a single witness. Lincoln rarely attacked a witness, believing the damage done to his relationship with the jury by that action might be more harmful than any gain from disputing facts. These were all friends and neighbors, good people, and it was taken for granted they were telling the truth. But even good people tended to remember things different, 'specially when there was a whole hullabaloo going around them. He pointed at the diagram that had been used by both sides to try to figure out who was standing where and doing what at what point. The jurors had heard a lot of testimony about where the various players were at each point in the event, who was to the east or the west or the northeast, but, and here he scratched his head, if they could figure out for sure what was goin' on each few seconds, maybe they could explain it to him. John Crafton remembered it one way, Mr. Short remembered it differently—and they were both in the middle of it. No question both of them were trying to tell the truth, but if they couldn't agree on how it happened, well, how could anybody expect the jurors to sort it out? He threw his arms into the air, palms up, his actions saying quite clearly that he had no answer to that question.

Unlike Broadwell and Logan, in his summation Lincoln made a point of acknowledging the prosecution's case, making a point to praise the fine work done by his old friend John Palmer. At that point he may have even mentioned, just as an aside, that there probably was no finer advocate in all of Illinois than John Palmer, a man who had trained so many fine lawyers, among them the very man sitting in judgment of this case, His Honor Edward Y. Rice.

Mr. Palmer had made some very strong points, he reckoned, and he wasn't about to dispute all of them. As he did in most of his cases, he reviewed the prosecution's case conceding certain elements of it: Quinn Harrison carried a knife and slashed Greek Crafton. But it wasn't necessary for him to cast doubt on the whole of the prosecution; while he mentioned most every witness the jurors had sweated through, his real focus was on those few issues that made all the difference in the outcome.

In this part of his argument he would speak without any great display of emotion, as much a professor leading his students through an especially interesting lesson as a lawyer fighting for his client's life. But slowly he wound his way into the core of the case, moving step-by-step away from the facts into his interpretation of them.

At some point in his presentation he paused to remove his coat, which he put on the back of his chair, maybe even for effect. He then began the next portion, finally letting loose the oratorical abilities that had marked him as a man to be watched and appreciated. A man who had the speaking gift, who through his words could spark true emotions in his listeners. This was the A. Lincoln that the spectators had packed the courtroom to hear. His voice until this moment had been pleasing as he lulled not just the jury, but everyone in the courtroom, into an area of comfort, and now with ease he took them out of it. The change in his voice and manner was dramatic. His close friend Joshua Speed once described this Lincoln: "Generally he was a very sad man, and his countenance indicated it. But when he warmed up all sadness vanished, his face was radiant and glowing, and almost gave expression to his thoughts before his tongue would utter them." His voice grew louder and was colored with emotion. His words cut through the calm he had created. Lincoln had laid the groundwork, he had done the legal job, now he was going for the heart.

In *Shaw v. Snow Brothers*, two underage boys had signed a contract to buy a team of oxen and a plow, then refused to pay for it, claiming the contract was invalid because of their age. The law was on their side, but Lincoln spoke for the plaintiff. "Gentlemen of the jury," he pleaded, "are you willing to allow these boys to begin life with this shame and disgrace attached to their character? If you are, I am not. The best judge of human character that ever wrote has left these immortal words for all of us to ponder: 'Good name in man and woman, dear my lord, Is the immediate jewel of their souls:

"'Who steals my purse steals trash, 'tis something, nothing, 'twas mine, 'tis his, and has been slave to thousands;

"'But he that filches from me my good name, Robs me of that which not enriches him And makes me poor indeed.'"

Without needing to retire for discussion, the jury in

that case found for Lincoln's client in the full amount. It was not uncommon for Lincoln to use the Bard to make a point; as he might say, he was raising up that common man to the best level of his understanding.

As he began an emotional appeal he took off his vest and his stock, he hooked his fingers in his suspenders and began telling his stories. As he built into his speech, though, one of the suspenders fell off his shoulder and hung loosely at his side. He made no move to pull it up, and it was questionable he even knew it had slipped, so totally engrossed was he in his remarks. This was the Lincoln that Hitt had seen on the platform with Douglas, the man with the ability to make every single person in a large crowd feel as though he was speaking directly to them. At times like this Lincoln might talk about impulsive young men ripe with emotions, emotions that sometimes get confused and wrapped together and cause people to overcome their own good sense. He might reach back into his own younger days and begin talking about his own errors. He had been pretty good with an ax in those days; he had won several rail-splittin' contests, and there was one mighty important lesson he had learned: once a rail was split there is no putting it back together.

There was no doubt at all in his mind that given another opportunity these two decent young men would have found another way to settle their differences; this outcome wasn't what they wanted, it wasn't what anyone wanted. And a lot of people were suffering because of it. But there was no going back; that rail had been split and this was what they were left to settle. The damage was done.

And then he started talking about Greek Crafton. In case anyone didn't know it, Greek had trained in his office. He was going to be a lawyer, maybe even standing right here defending another young man. Lincoln described Crafton, talking about his skills and the delights he took at small things. He talked about his own pain in accepting the fact that he was gone, that all that promise would never be fulfilled.

A sob came from the gallery. A family member, or a close friend. This is what that family had craved, somebody to stand up and talk about their loss. The fact that it was the lawyer for the accused made it all the more remarkable.

The coat that Greek Crafton had been wearing during the affray was lying on a table. Lincoln picked it up and held it open, the gash marks made by Harrison's cutting clearly visible. He shook his head so sadly at this evidence of the loss. Then he laid it carefully on the table and began quoting Shakespeare, who always seemed to have the right way of expressing his feelings.

"'But here I am to speak what I do know. You all did love him once, not without cause:

"'What cause withholds you then, to mourn for him? O judgment! thou art fled to brutish beasts, And men have lost their reason. Bear with me; My heart is in the coffin there with Caesar, And I must pause till it come back to me.'"

But, he continued, it is possible to mourn the death of a son, a loved one, a friend, without placing blame for it on a man innocent of all murderous intent. What happened was a tragedy, and to find Quinn Harrison guilty of anything other than being young and impetuous and scared would do nothing but further the tragedy. His words poured out of him without even a slight pause, and sounded to Hitt almost musical in their rhythm.

And then Hitt, who was once again captivated by Lincoln, suddenly realized why he was so deeply affected by this speech: Lincoln meant every word of it. This wasn't a performance, not the flimsy theater he had seen other lawyers put on for a jury. This was not an attempt to manipulate emotions; Lincoln's compassion was real, the tragedy was real to him.

His presentation stretched for almost two hours. His shirt was drenched with sweat, and he didn't hesitate for an instant. The court crier, Captain Kidd, later described the proceedings: "He closed by picturing to the jury the sad consequences of the homicide to his friend and neighbor, Harrison, and his son and the relatives of his former student, so vividly and full of feeling that one after another of those twelve different natures seemed to join with him in sympathy, the briny messengers stealing over the furrowed cheeks of the old and the health-flushed cheeks of the younger members of the jury, each foretelling in tearful signs the nature of their verdict.

"Burst after burst of eloquence followed, until not an eye could be seen without a glistening telltale standing at the portal of the soul, having been enticed there from the niche of sorrowful affection to witness the earnest gesture and listening to the burning, soul-stirring eloquence of Abraham Lincoln, when his judgment and moral nature dictated that his cause was just and right."

Upon finishing his remarks Lincoln paused one last time in front of Greek Crafton's slashed coat and stared at it, as if saying his final goodbye. Then he sat down and busied himself moving papers while those in that room absorbed the impact of his words. Logan reached behind Quinn Harrison and patted Lincoln on his shoulder.

After allowing a time for that, Judge Rice pointed at the prosecution table. "Mr. Palmer."

John Palmer stood. What happened next has been a matter of debate among historians. Although people who were in the courtroom would later write about it, others dismissed it as fanciful recollections. While the harshness of Palmer's response may be debatable, clearly Lincoln's emotional appeal had rankled him. As Stephen Douglas had learned, following a speech by Mr. Lincoln was a burdensome task. Palmer stunned the courtroom by responding to it: "Well, gentlemen," he said, "you have heard Mr. Lincoln.

'Honest Abe Lincoln' they call him, I believe. And I suppose you think you have heard the honest truth..."

As it became apparent that John Palmer was launching an attack on his opponent, there was not a murmur in the room.

"...or at least what Mr. Lincoln believes what he had told you to be the truth. I tell you, he believes no such thing!..."

Every eye was riveted on him. One of the boys sitting in the window whistled to those outside to hush. No one wanted to miss a word of this.

"...That frank, ingenuous face of his, as you are weak enough to suppose, those looks and tones of such unsophisticated simplicity, those appeals to your minds and consciences as sworn jurors are all assumed for the occasion, gentlemen. All a mask, gentlemen. You have been listening for the last hour to an actor, who knows well how to play the role of honest seeming, for effect."

He would have continued this attack on the fiber of Lincoln's being, but Lincoln finally had had enough. "Mr. Palmer!" he said, standing. Hitt noticed his fists were balled, tightly clenched; this was the type of charge that could destroy a man's career if it was to be believed. Palmer was calling him out as a charlatan. In a firm, but controlled voice, Lincoln responded, "You have known me for years, and you know that not a word of that language can be applied to me."

The two men glared at each other across the room for several seconds. They had been colleagues, friends, for decades. Each of them had great respect for the other. The irony that a trial about a cutting death that had started with an exchange of angry words had finally led to this was not lost on the participants. After several seconds, Palmer relaxed, his shoulders drooped, and he admitted, "Yes, Mr. Lincoln, I do know it. And I take it all back."

Shelby Cullom later remembered that the two men each moved forward a few paces and shook hands, defusing an ugly moment. And as they did, the courtroom erupted into applause. Minutes later John Palmer resumed his closing argument, perhaps having successfully planted some doubt in the minds of jurors about Lincoln's emotional appeal. There were a lot of folks who were wary about being fooled by lawyers at that time, as the methods employed by some lawyers were suspect. Even Lincoln had once written about the "vague popular belief that lawyers are necessarily dishonest..." especially in the matter of fees. "...Yet the impression is common, almost universal." Lincoln clearly had touched the jury with both his reasoning and his understanding of human nature. Palmer had no choice but to cut it down. He had apologized, but the residue of his attack could not be erased from jurors' minds.

The *State Journal* reported that Palmer "spoke with marked ability for three hours, evincing great ingenuity in handling the testimony, interspersing many remarks upon human nature and human passions, the duties of the citizens and the spirit of the law."

But of course, it was more than that. Palmer made an exceptional closing argument. He had a dazzling command of the facts and the law, and he diligently applied them to the evidence and the testimony the jurors had heard. He laid out a line of reasoning as straight and easy to follow as the railbed from Springfield to Chicago. But he was in the unfortunate position of following Abe Lincoln.

Lincoln had carried the spectators to an emotional pitch and he had held them there. The sight of the man holding up the cut coat of a young man he had tutored and bemoaning the senseless loss had broken hearts. He was not the only man in that courtroom crying, and when he was done the spectators were drained. Palmer had stepped into that moment. It was too much to expect him to be able to bring them back to that high point; even if it was possible, he wasn't that type of orator.

Hitt guessed that he knew it, too, which is why he had opened his summation by trying to challenge the truth of that emotion. On a factual basis his argument was the equal of Lincoln's, maybe even stronger. He might have even quoted Lincoln's own words: "I do mean to say that, although bad laws, if they exist, should be repealed as soon as possible, still while they continue in force…they should be religiously observed." The laws defining self-defense might not be to every man's liking, but the jurors were not allowed to make up their own interpretation of the law. And according to the law, this was not self-defense.

As Palmer continued, the heat in the courtroom, which somehow had seemed tolerable to Hitt during Lincoln's argument, had once again become oppressive, causing people to lose track of his argument. He had finished strong though, recalling the words of P. M. Carter telling Greek Crafton, "If you ever attack Quinn I believe he is determined to kill you," and Greek himself telling James Zane, "he threatened to kill me if I jumped him." And finally quoting William Purvines, who testified that if Crafton attacked him, "He would cut his guts out; he would as soon kill him as kill a dog." The prosecution concluded and now it was time for the jury to weigh in.

Judge Rice faced the twelve men and reviewed with them what they had to do. He laid down the law carefully, avoiding a sense that he favored one side or the other. As he did, Hitt once again considered the miracle that in such a brief time these twelve individuals had seemed to become one. Only a few days earlier several of them seemed to be reluctant participants in this jury, and now they were sitting there, most of them with their hands folded in their laps, ready, even seemingly eager, to take on this responsibility.

Before they were sent out to deliberate, Lincoln made a motion; it seemed to him, he said, that much of the case might be narrowed down to the testimony of Silas Livergood. Livergood had seen the fight, he had heard Harrison pleading he didn't want it. He had seen him holding on to the rail and being pulled by Mr. Short and the Crafton brothers. He had seen the cutting. Lincoln proposed to stipulate that Mr. Hitt should read all that witness's evidence to the jury before they retired.

This was a most unusual request. Hitt blushed bright red at the mention of his name, being unused to any mention in a trial. Judge Rice asked Palmer how he felt about that. Initially he said he didn't have a problem with it, but asked for a few minutes to confer with his fellows. When he spoke again, he had changed his thinking, he said. He didn't think it was a good idea at all and was against it.

Judge Rice accepted that, turning down Lincoln's motion. At eleven minutes after four o'clock the jury filed out, moving into their apartment, and once again the courtroom relaxed. There was a lot of discussion as to the merits of the case and, while Hitt could not confirm it, he suspected more than one wager was placed on the outcome. All the participants came together in the well of the courtroom; there was some hand-shaking and Lincoln and Palmer stood to the side in discussion, perhaps clearing up any sour feelings.

Rather than dispersing, the crowd mostly stayed put, an indication that they didn't expect the jury to stay out talking about it for too long. And they didn't want to miss it. A lot of people were lounging around the courthouse, some smoking, many chawing, a few tipping cider jars, talking about the case. From time to time voices got loud in dispute; there were as many of Greek's friends and supporters as there were Harrison people, and at least once peacemakers had to step between them to prevent blows.

The principals were not in sight; Harrison had been taken back into custody to await the verdict. Hitt wondered what it must feel like to be him; to know that twelve men are deciding your fate, whether you will spend the remainder of your life—however long that might be—as a free man or a prisoner. Most feelings he could at least imagine; not in this case though. He didn't know Harrison at all, and his demeanor during the trial had given away nothing; but if it were him, he would be terrified.

Although he did not prepare a transcription of the closing arguments, Hitt was still tasked with reporting the trial details to the newspapers. He began writing his notes; he had barely begun when Captain Kidd began spreading the word that the jury had reached its verdict. While juries rarely debated for substantial periods, they had been at it for only an hour and nine minutes, an unusually brief period. Whatever their verdict, they had come to agreement quickly.

The courtroom filled quickly, and even after it seemed to be crammed full, yet more people somehow managed to squeeze inside. After every inch of space was filled, additional people gathered below the windows, waiting for one of the boys sitting on the sills to shout out the verdict. Lincoln and Logan walked in on either side of Harrison. Hitt thought Harrison looked especially pale; his head was bowed and his eyes stared at the ground. The prosecution team hurried in as a group. And finally, Judge Rice took the bench and asked that the jury be brought in.

As the jury filed in, everyone looked for one of them to give away the verdict, but not one showed an indication of their decision. When they were seated, just as it had been for decades, the judge asked the foreman, "Has the jury reached a verdict?"

"We have, your honor," he said, handing the bailiff a folded slip of paper.

Judge Rice asked Quinn Harrison to stand. Lincoln and Logan stood with him. The judge unfolded the verdict. He was not a man for drama, so he read it aloud instantly. "'We, the jury, find the defendant *not guilty* as charged in the indictment.'" The courtroom exploded with jubilation and horror! The hurrahs were far louder than some jeers from Crafton's supporters. Seconds later a second cheer came from outside as the news spread. Judge Rice banged his gavel again and again trying to restore order, but few heeded him. Hitt turned in time to see Peyton Harrison hugging Lincoln; he was close enough to see tears running down Harrison's face. Logan and Quinn Harrison were standing behind the table, Logan with one hand on Quinn's shoulder, the two of them in a deep conversation. The prosecution team was gathering its papers, doing busywork, but their bowed heads made it impossible to see the expressions of their faces.

Hitt made a note to himself about the verdict, which he would later affix to the transcript. "The announcement was received by the large crowd in attendance with great and long continued applause."

After several minutes the courtroom calmed, and this time Rice was able to gavel it into silence. There was still work to be done; he asked Harrison to stand and then officially released him, telling him he was free to go and wished him Godspeed. Harrison shook hands in turn with Lincoln, Logan and Cullom. Judge Rice thanked the jury for their good work, praising them for giving their time, their patience and their thoughtful deliberations—although giving no hint if he was in accord with their decision—and then released them. With that he congratulated all the attorneys for their fine work in representing the interests of the state and the accused. Then he looked at the gallery and said, "It's time to go home, folks," and with a single gavel blow to the sound block ended the trial.

"Court is adjourned," Captain Kidd said firmly.

Within days newspapers had reported the verdict. One enterprising journalist quoted a young Springfield boy named William B. Thompson, who told him, "Mr. Lincoln saved Quinn Harrison, but it was a very hard fight. We boys followed it throughout. All of us who were able climbed to the windows. The others hung around the doors of the old courthouse. We listened with most careful attention to everything Lincoln said. His argument to the jury for Quinn Harrison made a lasting impression upon us. Harrison was acquitted. We boys agreed that Lincoln's speech and earnest manner did it, rather than the evidence."

It was already evening by the time Lincoln returned to his office. Although still enjoying the glow of his victory, he laid his tall silk hat on the top of his rolltop desk and set to work on several other pending matters, among them writing a check for $7.67 to Mr. W. P. McKinnie, a local farmer. He dealt with some correspondence that had been put aside the previous few days and then, as the night finally cooled, he went home. He would participate in several smaller trials in the next few months, before giving himself over completely to the business of running for president, but none of them would have the impact or engender as much attention as the murder trial of Peachy Quinn Harrison. The last great trial of Abraham Lincoln's legal career was done.

CHAPTER SIXTEEN

By the time Hitt had packed his supplies and readied to leave, the clamor had subsided. He had watched as Harrison had left the courtroom in a circle of his friends, who had remained respectfully low-key. After the commotion of the last few days, the echoes in the almost deserted courtroom were unsettling. Before departing, Hitt shook hands with Judge Rice, who praised his work and told him he was a welcome presence in his courtroom anytime, then each of the attorneys. Lincoln also thanked him and said he looked forward to working with him sometime in the future. Hitt almost wished him well in the upcoming election but thought it might make him uncomfortable and so said nothing about it.

The next train to Chicago was scheduled to depart the following morning, so he spent a final night at the Globe putting the finishing touches on his transcript.

He would be at the telegraph office when it opened the following morning and, thanks to the miracle of rapid rail transportation, back in Chicago by that evening.

The village of Pleasant Plains was ripped apart by the verdict.

While Harrison's supporters were pleased at what they considered a just outcome, Crafton's family and friends were furious at what they believed was a terrible injustice. Subsequently, probably to satisfy them, shop owner Ben Short was arrested and charged with being an accessory to the murder of Greek Crafton. His actions in trying to prevent the tragedy somehow were construed to have assisted Harrison in the outcome. It was a charge with weak underpinnings and eventually was dropped long before it could be brought to trial.

The tensions and even the outcome did not affect the deep friendship between Lincoln and John Palmer. In fact, less than two months after the end of the trial, Lincoln sent a note to Peachy Quinn Harrison asking him to support John Palmer in his quest to fill a seat in Congress left vacant by the death of four-term representative Thomas L. Harris. "I have no doubt that our friends are doing the best they can about the election," he wrote. "Still, you can do some more, if you will. A young man, before the enemy has learned to watch him, can do more than any other. Pitch in and try. Palmer is good and true, and deserves the best vote we can give him. If you can make your precinct 20 votes better than it was last we probably shall redeem the country. Try. Yours truly, A. Lincoln."

Perhaps Lincoln hoped Harrison might rally those friends of his still angry at the prosecutor for his work in the trial. Lincoln's efforts in support of the Republican candidate failed, and Palmer was defeated by John A. McClernand.

Still, within months Palmer was able to return that favor. While Lincoln remained a dark horse candidate for the 1860 Republican presidential nomination, his victory in this trial contributed to his increased visibility. Freed, mostly, from the "Press of business in the courts," as he described it, he began speaking throughout the Midwest, seeking to establish that region as his base to support him during the convention. He was testing the political waters, while at the same time denying interest in the presidency, writing to a supporter, "In regard to the matter you spoke of. I beg you will not give it further mention. Seriously, I do not think I am fit for the presidency."

Only weeks after the conclusion of the trial, he received an invitation to speak at New York City's Cooper Institute the following February. His speech there was widely reprinted in its entirety. Publisher

Horace Greeley wrote in his *New York Tribune*, "No man ever made such an impression on his first appeal to a New York audience."

Lincoln had no organization, no campaign funds and little political experience. His opposition was Senator William Seward, who had all of these things. At the Republican convention in May, Judge David Davis, Stephen Logan and John Palmer, working together, helped secure the nomination for him. But it was another member of the prosecution team who claimed credit for it.

As a member of the Illinois delegation, which was hosting the Republican convention in Chicago, young Norman Broadwell was assigned the oft-thankless task of arranging the seating of the delegations. As he explained to his son, Rufus, he put the state delegations committed or leaning to Seward at the front, delegations favoring Lincoln in the center and whenever possible those committed to other candidates or in doubt in the rearmost seats. Seward led after the first ballot, but his supporters were separated from those in the back who had voted for a third candidate, while the Lincoln people moved easily among them. By the third ballot his missionaries had done their work, and thanks at least in part to Broadwell's seating chart, Lincoln had captured the nomination.

Lincoln was not in Chicago when he was nominated; instead he had spent the day throwing a ball around with his sons, then sitting in the telegraph office awaiting the results. The following November he was elected the sixteenth president of the United States. He guided this nation through the most devastating war in our history and survived only long enough to taste a great victory. He had held the Union together through the worst times and cemented his position as one of America's greatest leaders. For more than 150 years his life and his deeds have represented the best of this nation.

But, many other participants of the trial also went on to lead long and distinguished lives. John Palmer rose to the rank of brigadier general of the Union army in the Civil War and afterward served as military governor of Kentucky where, as promised, he drove "the last nail in the coffin" of "that abominable institution." Following the war, he was elected governor of Illinois and in 1890 was sent by that state to the U.S. Senate. Two years later he was considered a serious candidate for the Democratic presidential nomination. He did make a run for it in 1896 but at seventy-nine years old really had no chance.

The wealthy Stephen Logan offered substantial financial support to Lincoln's presidential run, but after being asked by President Lincoln to serve as a member of an 1861 delegation to an unsuccessful peace conference with Confederate representatives, he essentially retired. As a speaker at an 1865 memorial for his former partner after Lincoln's assassination, Logan lauded him as a man who "when he believed his client was right, especially in difficult and complicated cases, he was the strongest and most comprehensive reasoner and lawyer he ever met—or if the case was somewhat doubtful but could be decided either way without violating any just, equitable or moral principle, he was very strong—but if he thought his client was wrong he would make very little effort."

Lincoln's law partner, William Herndon, left the small sign announcing the firm Lincoln and Herndon hanging outside their office when his partner assumed the presidency, as Lincoln had requested. Herndon continued the practice of law after the assassination, with some moderate success, but eventually published a long and controversial biography of Lincoln that remains in print today.

The youngest member of the defense team, Shelby Cullom, went on to have a distinguished political career; in addition to serving as governor of Illinois, he was elected to three terms in the House of Representatives and five terms as a United States senator from that state. As a senator he championed significant legislation, including the *Interstate Commerce Act* of 1887, which guaranteed fair treatment to all customers of America's railroads. John Alexander McClernand, another member of the prosecution, served in the Civil War as a brigadier general. He saw action in several of the great battles of the war, including Shiloh and the siege of Vicksburg, but his contentious relationship with General U. S. Grant eventually caused him to resign from the army.

After a long and distinguished judicial career, Judge Edward Y. Rice served as a delegate to the state constitution convention and in 1870 was elected to Congress. He served one term and, after being defeated for reelection, returned to the practice of law. The Reverend Peter Cartwright continued with his popular ministry, and perhaps due to Lincoln's stalwart work in his grandson's trial, came full circle in his appreciation of the man. At an 1862 dinner with the president's political opponents he told them, "Once we (Lincoln and Cartwright) were opposing candidates for a seat in Congress, and, measured up in the ballot-box, I went down in defeat. But it was defeat by a gentleman and a patriot. I stand here tonight to commend to you the Christian character, sterling integrity, and far-seeing sagacity of the President of the United States."

Dr. John Million eventually became one of Springfield's leading citizens. Although it was not mentioned during the trial, it appears that during this time he was courting Mary Crafton, Greek's sister, whom he eventually married.

Given a chance to live a fulfilling life, Peachy Harrison could never seem to find satisfaction. In 1867 he married Emeline LaMothe Guillet, the widow of a Confederate officer. But his volatile personality caused a rift, and he left their marriage bed for years at a time, often disappearing into the untamed west. They had two children, but he became estranged from them, as well. Ironically, while he was absent, his younger brother Peter shot and killed a local farmer named William Kelly after a long feud. Peter fled to

Texas where he was captured and returned to Springfield for trial—and like his older brother was acquitted by a jury.

Peachy made news once again in 1885, when a dispute with his sister Sarah over their father's missing will became violent. He called her "a ghoul" for her actions at this time of grief. When she challenged him, he pushed her backward and she fell over a couch, breaking a rib. The Springfield newspapers claimed she had been thrown violently and was near death, using the opportunity to reprise the story of Abe Lincoln's last great criminal trial. Sarah made a rapid and full recovery while Peachy became known for the fanciful stories he told about his adventures in the old west and the time Lincoln saved his life.

Among the most successful participants in the trial in later years was, ironically, the steno man, Robert Roberts Hitt. As was noted in the *Congressional Record* honoring his long service, "All during the Civil War and later, Mr. Hitt was employed in many confidential capacities and his abilities and proficiencies were so well recognized that his services were constantly sought by commissions, by committees of Congress, by military courts and by the Executive departments." After the end of that war he accompanied General Grant on a notable world tour. In 1874 President Grant appointed him First Secretary of the American Legation in Paris, during which time he also served as Chargé d'Affaires. Returning to this country, he was appointed Assistant Secretary

"THE HONORABLE R.R. HITT, ILLINOIS," HALFTONE PRINT OF A DRAWING, GEORGE WILLIAM BRECK, HARPER'S WEEKLY, MAY 5, 1894, COLLECTION OF THE OF U.S. HOUSE OF REPRESENTATIVES

This drawing of the Honorable Robert Roberts Hitt appeared in Harper's Weekly *in 1894, long after his amazing transcription had led to him becoming a key aide to President Ulysses S. Grant and eventually a beloved twelve-term member of the House of Representatives.*

of State under James G. Blaine during the administrations of President James A. Garfield and President Chester A. Arthur. He was elected to the first of his twelve terms in Congress in 1882, eventually becoming Chairman of the Committee on Foreign Affairs. Among his many contributions was his bold stand against the *Chinese Exclusion Act*, arguing, "Never before in a free country was there such a system of tagging a man, like a dog to be caught by the police and examined, and if his tag or collar is not all right, taken to the pound or drowned and shot. Never before was it applied by a free people to a human being, with the exception (which we can never refer to with pride) of the sad days of slavery..." He served as Regent of the Smithsonian Institution and was elected a member of the National Geographic Society. In 1906 he was mentioned as a possible candidate for the vice presidency, but died in September of that year.

At some point Hitt's original transcript of the trial was bound with a ribbon and put aside, only to be discovered in 1989 in a shoebox stored in a garage of the Fresno, California, home once owned by Quinn Harrison's great-grandson. And from that meticulous transcript so perfectly stored over the years, we are left with the final direct link to the last great trial of Abraham Lincoln's legal career, an event that helped propel him to the presidency.

* * * * *

BIBLIOGRAPHY

We consulted numerous resources in our research, but those that repeatedly proved to be valuable are included here.

"Abraham Lincoln and Springfield." *Abraham Lincoln's Classroom.* http://www.abrahamlincolnsclassroom.org/abraham-lincoln-state-by-state/abraham-lincoln-and-springfield.

"Abraham Lincoln and the Law." *Abraham Lincoln's Classroom.* http://www.abrahamlincolnsclassroom.org/abraham-lincoln-in-depth/abraham-lincoln-and-the-law.

Angle, Paul M. *"Here I Have Lived": A History of Lincoln's Springfield, 1821–1865.* Chap. 6, "Social and Cultural

Growth." Springfield, IL: Abraham Lincoln Association, 1935. Accessed from the University of Michigan Library Digital Collections, http://quod.lib.umich.edu/l/lincoln2/0566798.0001.001/1:11.9?rgn=div2;view=fulltext.

Angle, Paul M., ed. *The Lincoln Reader.* New Brunswick, NJ: Rutgers University Press, 1947.

Barrett, Joseph H. *Life of Abraham Lincoln: presenting his early history...and a history of his eventful administration, and of the scenes attendant upon his tragic and lamented demise.* Cincinnati, OH: Moore, Wilstach & Baldwin, 1865.

Blackstone, Sir William. *Commentaries on the Laws of England in Four Books, Book I: The Rights of Persons.* Oxford, England: originally published by Clarendon Press, 1765.

Blackstone, Sir William. *Commentaries on the Laws of England in Four Books, Book II: Of the Rights of Things.* Oxford, England: originally published by Clarendon Press, 1765.

Britt, Robert Roy. "150 Years Ago: The Worst Solar Storm Ever." *Space.com*, September 2, 2009. https://www.space.com/7224-150-years-worst-solar-storm.html.

Dekle, Sr., George R. "Lincoln's Last Murder Case: The News Reports." *Abraham Lincoln's Almanac Trial*, November 20, 2015. http://almanac-trial.blogspot.com/2015/11/lincolns-last-murder-case.html.

Dekle, Sr., George R. *Prairie Defender: The Murder Trials of Abraham Lincoln.* Carbondale: Southern Illinois University Press, 2017.

Dirck, Brian. *Lincoln the Lawyer.* Urbana: University of Illinois Press, 2007.

Donald, David Herbert. *Lincoln.* New York: Touchstone, 1995.

Duff, John J. *A. Lincoln: Prairie Lawyer.* New York: Rinehart, 1960.

Ecelbarger, Gary. *The Great Comeback: How Abraham Lincoln Beat the Odds to Win the 1860 Republican Nomination.* New York: Thomas Dunne Books, 2008.

Emsley, Clive, Tim Hitchcock and Robert Shoemaker. "Crime and Justice—Trial Procedures." *Old Bailey Proceedings Online.* Version 7.0. https://www.oldbaileyonline.org/static/Trial-procedures.jsp.

Gridley, J.N. "Abraham Lincoln's Defense of Duff Armstrong. The Story of the trial and the use of the almanac." *Journal of the Illinois State Historical Society* 3 (1910): 24–45.

Harper's Weekly, August and September, 1859, et al.

Herndon, William H., and Jesse William Weik. *Herndon's Lincoln: The True Story of a Great Life.* Chicago: Belford, Clarke, 1889.

"History of Trial by Jury." *English Legal History*, June 10, 2013. https://englishlegalhistory.wordpress.com/2013/06/10/history-of-trial-by-jury.

Koerner, Brendan. "Where Did We Get Our Oath?" *Slate*,

April 30, 2004. http://www.slate.com/articles/news_and_politics/explainer/2004/04/where_did_we_get_our_oath.html.

Langbein, John H. "The Criminal Trial Before the Lawyers." *University of Chicago Law Review* 45 (Winter 1978): 263–316. Accessed from http://digitalcommons.law.yale.edu/cgi/viewcontent.cgi?article=1547&context=fss_papers.

The Law Practice of Abraham Lincoln, 2nd ed. http://www.lawpracticeofabrahamlincoln.org/Search.aspx.

"The Lawyers: Stephen Trigg Logan." *Mr. Lincoln & Friends.* The Lehrman Institute. http://www.mrlincolnandfriends.org/the-lawyers/stephen-trigg-logan.

Leidner, Gordon. *Lincoln's Gift: How Humor Shaped Lincoln's Life & Legacy.* Naperville, IL: Cumberland House, 2015.

Lewin, Travis H.D. "Lincoln's Last Murder Trial: People v. Harrison." June 14, 2014. http://apps.americanbar.org/litigation/committees/trialevidence/articles/spring2014-0614-lincolns-last-murder-trial-people-v-harrison.html.

The Lincoln Log: A Daily Chronology of the Life of Abraham Lincoln. http://www.thelincolnlog.org.

"Lincoln Neighborhood." *National Park Service*. https://www.nps.gov/liho/lincoln-neighborhood.htm.

McDermott, Stacy Pratt. *The Jury in Lincoln's America.* Athens: Ohio University Press, 2012.

McGinty, Brian. *Lincoln's Greatest Case: The River, the Bridge, and the Making of America.* New York: Liveright Publishing Corporation, 2015.

"The Preachers: Peter Cartwright (1785–1872)." *Mr. Lincoln & Friends.* The Lehrman Institute. http://www.mrlincolnandfriends.org/the-preachers/peter-cartwright.

Wilson, Rufus Rockwell. *Intimate Memories of Lincoln.* Elmira, NY: Primavera Press, 1945. Accessed from the Internet Archive. https://archive.org/details/intimatememories00wils.

Rothschild, Alonzo, and John Rothschild. *"Honest Abe": A Study in Integrity Based on the Early Life of Abraham Lincoln.* Boston: Houghton Mifflin, 1917.

The story of the Sangamon County court house. Springfield, Ill., Phillips bros, 1901. Accessed from the Library of Congress, https://lccn.loc.gov/24005641.

"The strongest man in the world," *Chicago Tribune*, June 6, 1859. Accessed from http://archives.chicagotribune.com/1859/06/06/page/2/article/the-strongest-man-in-the-world-prostrated.

von Moschzisker, Robert. "The Historic Origin of Trial by Jury," *University of Pennsylvania Law Review* 70 (November 1921): 1–13. History.com/This Day in History.

Wallace, Joseph. *Past and Present of the City of Springfield and Sangamon County, Illinois, Volume 1.* Chicago: S. J. Clarke Publishing Company, 1904.

White, Jonathan W. *Lincoln on Law, Leadership and Life.* Naperville, IL: Cumberland House, 2015.

Wilson, Rufus Rockwell. *Intimate Memories of Lincoln.* Elmira, NY: Primavera Press, 1945.

ACKNOWLEDGMENTS

First and foremost I would like to thank my co-author, David Fisher. David is not only an incredibly gifted writer but he approached this project with a laser focus on, and appreciation for, historical detail that helped tell the story surrounding the trial. After all, when one ventures to chronicle anything Lincoln related, you best be prepared to present something beyond the hundreds of deeply researched accounts that already exist. Thanks to David, we were able to bring to life even the more obscure characters as well as Springfield, Illinois, in the year before it would become defined by the 16th President. The recurring question I would send as we exchanged notes: "How do we know this to be true?" Each and every time he would send back an historical reference and/or proof that unquestionably supported the contention. I must admit that I was thoroughly impressed when he was able to present a picture of Captain T. S. Kidd, the court crier, to back up his physical description.

I should also note that David brought this potential partnership to me along with his talented literary agent, Frank Weimann. David and I did not know one another before this project and a collaboration such as this can end up being fraught with potential personal and/or professional mines. We encountered none. At each milestone in the process, David overdelivered and I feel truly lucky to have him as my partner in this and a future endeavor.

I also want to thank Peter Joseph from Hanover Square Press who was willing to bet on this as one of his first releases with a new imprint. His notes and feedback were always helpful and on point. Simply put, his intellect and appreciation for the subject matter just made the book better.

My sister, Ronnie, a federal judge, and her husband, Greg, always serve as models of rectitude in the law and beyond for me and I'm convinced Lincoln would have admired both of them for it. I have long been so proud of their insightful daughters, Dylan, Teddy and Finn, who are always available to offer me more than a dollop of humility.

No one has unequivocally supported me in the myriad of sometimes surprising life and professional choices more than my loving mother ("eema"), Efrat, who helped shape the person I am today.

My father, Floyd, to whom this book is dedicated, remains the most trusted advisor in my life. On anything from an opinion article I am contemplating, to a house I am buying, to a legal issue I am analyzing, to this book, I always run it by Dad for his thoughts first. I recall as a teen asking how I might become a better writer, like him. "There is only one way," he responded, "read more books." While I am not certain I heeded the advice immediately, those words continue to inspire.

Finally I want to thank the phenomenal woman I have been lucky enough to partner with on my most important project. Thank you, Florinka, for bringing our child into the world with me and for surpassing even the unreasonable hopes and dreams I had for you as a mother. I am truly blessed to have you in my life. Everything else pales in comparison to the love I have for my beautiful son, Everett. Becoming a parent, of course, changes everything.

—*Dan Abrams*

I would like to begin by acknowledging our publisher, Peter Joseph; it has been my great pleasure watching his professional success and Dan Abrams and I are thrilled to be part of his first list. I also want to thank Dan, who has been the complete collaborator, contributing to every aspect of this book with his creativity, curiosity, intelligence and insistence on meeting and maintaining the highest standards. He pushed and pulled in all the right places and did so with absolute professionalism. I also appreciate the efforts of Fred Rappoport, who brought us together, and our agent, Frank Weimann of Folio, who always has my back.

Dan and I would also like to thank Maureen Wilburn, whose research concerning her family, the Harrisons, has proved invaluable. We urge interested readers to visit www.ourbigfamilyhistory.com for the 330-year story of survival, endurance, and love of family and country.

We would also like to thank Donna Aschenbrenner of Donna's House of Type, Inc. of Springfield, Illinois, (217-522-5050) for supplying the Lloyd Ostendorf Images, and Curtis Mann, Sangamon Valley Collection Manager at the Lincoln Library.

I was very fortunate at Syracuse University to have been a student of the late Dr. Michael Sawyer, whose love of constitutional law was infectious and who played an important role in my life. And then I had the privilege of working with Johnnie Cochran, who never stopped loving the possibilities of our legal system. My brother and sister-in-law, Richard and Elise Langsam, and my nephew Andrew Glenn are all attorneys and represent the best of that profession. I also have several friends, lawyers all, whom I greatly admire. These include the inimitable Captain Arthur Perschetz; George Zelma, whose impact on my entire family will never be forgotten; Arthur

Aidala, who is always there on the other end of the phone when needed; Saul Wolfe and the twins, David and Jon, who fight to make this profession better; Paul Reichler, who has spent his life using international law to fight for a better life for people around the world; Mike Vecchione, who has fought all the legal wars with tremendous integrity; David Stein, who has used the law to protect the rights of working Americans; Keith Stein, a friend of both Dan and mine; my neighbors Victor Kovner, Marty Shaw and the late Professor Jerry Leitner, and the kite-flying Jon Lindsey who moves all the pieces on the board with such joy and skill. I want to make a special mention of the late Judge Stanley Sklar, whose love for the law was so complete that on the morning of 9/11, while New Yorkers were fleeing north from Ground Zero, Stanley was going south, into the chaos, believing there might be a need for a judge.

My friend Brian McLane is not a lawyer, but he has spent his life skillfully using the law to make life better and easier for those people who need it most. Brian has never hesitated to go after the most powerful people on behalf of the less powerful, and most often has succeeded. On the list of people I most admire, you will find his name at the top.

I also want to offer my gratitude to James M. Cornelius, Ph.D., the Curator of the Lincoln Collection at the Abraham Lincoln Presidential Library and Museum. From the very beginning of this project he has offered his support and guidance and his continued support is greatly appreciated.

Finally, I am so fortunate to have a partner who somehow manages to find the right words and gestures at the right time, every time. My wife, Laura, makes my life immeasurably better every day. As I often tell her, I am a very lucky man to have met her and that she chose me.

—David Fisher

INDEX

Page numbers of illustrations and their captions appear in italics.

Abraham Lincoln
Library and Museum,
Springfield, Ill., 15
Adams, John, 206
as defense lawyer,
Boston Massacre,
207–9
fate of Private Killroy
and, 209
*The Works of John
Adams Esq, Second
President of the
United States*, 207
African Americans, at
Harrison trial, 90
See also slavery
Allen, Charles, 68–69
Allen, John, 250–51
relationship with
Lincoln, 250–51
trial testimony, 250–51,
257–59, 301–7

American Revolution, 142, 327
American South, 181, 226
See also slavery
Anderson, George, 64–65
Anderson, Jane, 64–65
Aristotle, 145
Armstrong, Duff, 66–69, 112, 195, 328
Armstrong, Jack and Hannah, 67, 69, 148
Arnold, Isaac, 56
Arthur, Chester A., 371
Astor House, N.Y., 296
Atherton, Albert, 262–67
Austin, Benjamin, 199
Austin, Charles, 199

Backwoods Preacher, The (Cartwright), 98
Baker, Edward D., 49–50, 62
Ballou's Pictorial Drawing-Room Companion (magazine), 141
Bates, Edward, 118
Beardstown, Ill., 67
Beck, Sarah, 218, 219
Beckwith, Hiram W., 123
Black Hawk War, 28, 49, 53, *100*
Blackstone, Sir William, 48, 145, 195
  on a dying declaration, 214–15
  on self-defense, 195–96
Blaine, James G., 371
Bone, John C., 181–84
Boston Massacre, 207–9
  Adams as defense lawyer, 207–9
  deathbed declaration of Carr, 215
  fate of Private Killroy, 209
  law of self-defense and, 207–9
Bouvier's Law Dictionary, 307
Bramwell, Zenas, 88
Breese, Sidney, 80
Broadwell, Norman M., 91, 107, 154, 204–5, 333–34
  admission of hearsay evidence and, 291–92
  closing arguments and, 333–35, 342, 347
  law offices of, *78*
  on Lincoln's persuasiveness, 206
  Lincoln's presidential candidacy and, *78*, 366
  on Lincoln's self-defense strategy, 206–8
  questioning of Livergood and, 136–37
Broadwell, Rufus, 366
Brown, Ben, 88
Brown, John, 117
Brown, Thomas Jefferson, 223, 225

Brundage, James A., 92–93
Butterfield Overland Company, 220

Cameron, Simon, 118
Camp, William, 93
Carter, P. M., 293–95
Cartwright, Madison, 292–93
Cartwright, Nancy Purvines, 292
Cartwright, Rev. Peter, 31–32, 33–34, 82, 97–99, 212–13, 267, 271–72, *278*, 316
  The Backwoods Preacher, 98
  Lincoln and, 33–34, 98–99, 103, 109, 268–69, 273
  post-trial life, 369
  prosecution's strategy regarding his testimony, 212–16
  reevaluation of Lincoln by, 369
  slavery issue and, 98, *278*
  trial testimony/Crafton's deathbed declarations and, 40, 81, 109, 119–20, 269, 273–82, 288–90
Cass, Bob, 88, 92
Cassell, Robert, 29
Catron, John, 202
Chase, Salmon P., 97, 118, 222
Chicago Daily Press, 18
Chicago Press and Tribune, 18, 19
Chitty, Joseph, 48
Choate, Rufus, 324
Clinton Daily Public, 192–93
Cobham, Lord, 253
Commentaries on Equity Jurisprudence (Story), 48
Commentaries on the Laws of England (Blackstone), 48, 145, 195, 214–15
Commonwealth v. Selfridge, 199–200
Continental Congress (1774), 142–43
Cooper Union, Lincoln's address at, 365–66
Cornelius, James M., 15
Corwin, Thomas, 194, 324
Couldock, Charles Walter, 332
Crafton, Edmund, 184–86
Crafton, Elizabeth Catherine Harrison "Eliza," 35, 104, 311
Crafton, Greek, 79–80, 107
  character and personality, 83, 201
  deathbed declarations, 31–32, 34–35, 81, 213–15, 275–76,

280–81, 289–90, 297–99, 317–20
events leading to his murder, 35–37, 104–6
fight with Harrison and mortal wounding of, 30–31, 104, 125–39, 234–48, 258–59
as larger, stronger than Harrison, 31, 80, 103, 157, 198, 240, 261, 265–66, 300, 337
legal training with Lincoln, 33, 103, 165, 201, 351
threats by, 36, 37, 80, 104, 108–9, 157, 185, 197, 210, 244–45, 251, 257, 260, 291–92, 300, 308–9, 312, 334, 338, 341
Crafton, John, 30–31, 37, 103, 104, 125–26, 154–55, 158
Grand Jury hearing and, 107, 166
Lincoln's cross-examination of, 165–71
role in the fight with Harrison, 127, 131, 133–34, 138, 235–39, 246, 248, 258–59
testimony at trial, 140, 158–71, 219, 347
Crafton, Wiley and Agnes, 159, 260
Crafton, William, 35, 104, 311
Crystal Palace, London, 296
Cullom, Shelby Moore, 39, 335–37
as assistant for the defense, 39, 88, 90–91, 340–41, 342, 355
closing argument and breakdown, 335–37
Palmer's attack on Lincoln and, 355
post-trial life and accomplishments, 368
safety measures for Harrison and, 102

Danville *Illinois Citizen*, 56, 57
Davis, David, 57, 72, 324–25, 366
Day, Josiah, 193
Declaration of Independence, 143
Democratic Party/ Democrats, 22, 24, 34, 49–50, 51, 72, 74, 98, 147, 367
Dickens, Charles, *Oliver Twist*, 283
Donati's Comet, 223
Donner-Reed Party, 146–47

Douglas, Stephen, 18, 22, 50
Lincoln-Douglas debates, 13, 18, *77*, 147, 224–25, 350
praise of Lincoln's skill with juries, 324
Senate race against Lincoln, 22, 147
slavery issue and, 19, 222
Douglass, Frederick, 179–80
My Bondage and My Freedom, 179

Earley, Jacob M., 50–53
Edwards, Ninian W., *153*, *155*
Effie Afton trial, 18–19, 59, 328–30
Epler, Jacob, 216, 304, 306, 315–19

Fanes, James, 279
Fillmore, Millard, 24
Fink, Mike, 33
Fisher, Archibald, 61–62
Fleming, Samuel G., 327
Fleming v. Rogers & Crothers, 327–28
Fraim, William Fielding, 63

Galloway, Samuel, 96–97
Garfield, James A., 371
Gatton, Josephus, 87
George III, King of England, 143
Goings, Melissa, 29–30
Grainger v. State, 202
Grant, Ulysses S., 370–71, *371*
Gravelet, Jean-François, 156
Gray, Samuel, 209
Greeley, Horace, 19, 366
Green, Bowling, 47
Greenleaf, Simon, 48

Harnett, Daniel, 171–73
Harper's Weekly, legal wit article, 113–14
Harris, Thomas L., 364
Harrison, Emeline LaMothe Guillet, 369
Harrison, Frances, 186
Harrison, George M., 99, *100*
Harrison, "Peachy" Quinn, 22, 82, 83, 103, 337
arrest and charge of murder, 32, 41
in the courtroom, 82, 147, 154, 186, 190, 233, 244, 264–65, 282, 287, 298, 359–60
Grand Jury indictment of, 41, 106, 110, 111,
"Have I no friends here" plea, 173, 203, 338
health of, 261–67, 274–75, 337

in hiding, 37, 102, 336
initial interview with Lincoln, 102–6
jury deliberation and, 358–59
knife (murder weapon) and, 31, 41, 80, 104–5, 109, 121, 122–23, 128–33, 157, 161–62, 164, 169, 172, 182–83, 189, 205, 236, 247, 248, 252, 259, 260, 294, 300, 312, 334, 341, 348
Lincoln and Logan hired to defend, 38–39, *38*, 96, 211
Lincoln requests his support for Palmer's Congressional run, 364–65
murder case, general facts, 30–32, 35–36, 37
post-trial life, 364, 369–70
prohibited from testifying, 198, 312, 339
self-defense argument and, 44, 80, 130, 133–38, 156–57, 169, 176, 183–84, 196–97, 205–6, 209, 238–39, 245–48, 252–54, 261, 288, 291, 300–315
size of, 30–31, 80, 82, 103, 157, 198, 240, 261, 300, 337
verdict and, 360
witnesses testify about the fight and, 125–39, 158–78, 234–52, 258–59
Harrison, Peter, 36, 174, 308, 369–70
Harrison, Peyton, 33, 38–39, 187, 190–91, 263, 360
Harrison, William, 186
Hay, John Milton, 22
hearsay evidence, 34–35
deathbed declarations, 34–35, 213–15, 275–76, 282–89, 317
threats by Crafton, 210, 251–57, 291–92
trial of Sir Walter Raleigh and, 253
Henry, Frederick, 173–78
Henry II, King of England, 144
Herndon, Archie, 152, *155*
Herndon, William, 55, 66, 91, 152, *155*, *192*, 193, 201, 334
biography of Lincoln, *192*
as co-counsel, Harrison trial, 39
on Lincoln as defense lawyer, 85
on Lincoln's beliefs, 201

as Lincoln's law partner, 26, 39, 53, 148, 191, *192*, 194, 344
on Lincoln's outburst about admission of a dying declaration, 285–86
on Lincoln's style, 346
post-trial life and accomplishments, 368
Herndon's Lincoln (Herndon), *192*
Hill, William P., 250–51
Hinchey, William, 90–91
Hitt, R. Roberts, 13, 17–22, 39, 43, 221, *371*
aftermath of the trial and, 363–65
Allen testimony and, 302, 305–6
Cartwright testimony and, 280, 290
Chinese Exclusion Act opposed by, 371–72
Civil War and, 370
coroner's inquest and, 40
courtroom encounter with hostile men, 149–52
on courtroom layout, 89–90
courtroom temperature and, 178, 332
Crafton testimony and, 160, 162, 164, 165, 168, 171
discussion of the trial by, 217, 219–20
Effie Afton trial and, 18–19, 59, 328–30
fees, 20
Harrison trial opens, 71–72, 76
Henry testimony and, 178
on history of the courts system, 142–45
importance of his work, 321, 340
jury selection and, 83–84, 87–88, 90–92
Lincoln and, 18–20, 27–28, 30, 71, 114–15, 181, 222–25, 323, 328, 363–64
on Lincoln's argument for admission of dying declaration, 282, 284, 286, 287
on Lincoln's reputation, 323–29
on Lincoln's speaking ability, 350, 352
on Lincoln's wit, 114–15
Livergood testimony and, 127, 128, 132, 135, 139, 357
Million testimony and, 120, 122, 123
note about the verdict, 360–61

Nottingham testimony and, 311
as official scribe, Illinois State Senate, 148
opening or closing arguments and, 333, 340
Palmer's attack on Lincoln and, 353–55
as pioneer in stenography, 18
post-trial life and accomplishments, 370–72, *371*
Purvines testimony and, 309
reading material for, 141–42
second day of testimony and, 233–34
shorthand (phonography) and,16, 18
Short testimony and, 238, 239
slavery issue and, 180–81
in Springfield, 17, 25–26, 40, 44, 112, 191, 216–19, 321, 363
as stenographer, 17–20, 21–22, 71, 112, 123, 148–49, 155, 157
as stenographer, physical toll of, 187, 246, 248, 303
as stenographer, questions faced by, 123, 162, 172, 174, 175, 184, 258, 260
tools of his trade, 21, 93, 134–35, 141, 190
transcription of *The State of Illinois v. "Peachy" Quinn Harrison*, 15–16, 20, 372
transcription of legal discussion and, 252–57
transcriptions of the Lincoln-Douglas debates, 19–20, 147, 223, 225
transmission of proceedings via telegraph, 230–31, 321, 332, 364
trial observations by, 82, 158, 160, 168–71, 176, 178, 181–86, 188–91, 238, 248–49, 254, 255, 264–65, 269, 272–73, 274, 282, 334, 339
work habits, 172, 180, 187, 190, 216–17, 230, 277
"work sense," 164
Zane testimony and, 313, 314–15
Hobbes, Thomas, *Leviathan*, 196

Iouston, Sam, 181
Iutchinson, John, 193

'llinois State Journal, 32, 51, 107, 109, 231, 290
hires Hitt to cover the Harrison trial, 20
on Lincoln's voice, 344
'llinois State Register, 32, 110, 341–42
rwin, Robert, 165

ackson, Andrew, 272
enkins, Jameson, 24
ohn, King of England, 144
uries/jury system, 82–83
history of the courts system and, 144–45
Illinois law on juries, 83
Lincoln's skill with, 56–57, 84–85, 323–31
questions from, during testimony, 182
trial by jury as fundamental right, 142–43
selection for Harrison trial, 83–93
Twain on, 83

Kidd, Thomas Winfield Scott, 82, 145–48, 153, 157, 282, 352, 359, 361
on Lincoln's outburst about admission of a dying declaration, 285
Killroy, Matthew, 209
King, John, 255

law and courts in America, 14
accused prohibited from testifying on his own behalf, 198, 312, 339
admission of a dying declaration, 213–15, 275–76, 282–89
British law and, 46, 207
circuit riding and, 24–25, 229
closing argument, or summation, 330–31
courtroom layout, 89–90
cross-examination, prime rule of, 170
"double-teaming," 132
in the 1800s, 46
evolution of, 339–40
expert witnesses, 263
fundamental principles, 142
hearsay evidence and, 34–35, 210, 214, 251–57, 282–89
history of the courts system, 144–45
importance of lawyers in, 58
insanity defense and, 65–66

judicial robes, 228–29
juries and jury trials, 82–83, 142–43, 288
legal education and training, 47–49
legal wit, 113–14
Lincoln and, 43–44, 58, 59–60, 61, 94–95
murder trials as entertainment, 64, 143
oath taken by witnesses, 307
principles of, 46
rules of evidence, 46, 198
self-defense law, 14, 32, 80–81, 195–200
specialty in areas of law, 25
trial lawyers, storytelling skills of, 325
trial spectators, 76, 90, 93, 139, 149
trial transcripts and courtroom stenography, 21, 123, 340
truth as a defense against slander, 59
in the west, 49
witnesses, history of sequestration, 119
witnesses, letting have their say, 108, 280

Ledbetter, J. T., 162, 174, 219
Leviathan (Hobbes), 196
Lincoln, Abraham, *326*
appearance, 50, 54, 57–58, 73, 88, 96, 102
belief in predestination, 201
Black Hawk War and, 28, 49, 53, *100*
character and personality, 52–55, 88–89, 224–25, 254–55, 284, 349
Cooper Union address, 365–66
duel with James Shields, 36–37
elected president, 367
ethics of, 44, 225
Hitt and, 18, 20, 27–28, 30, 71, 114–15, 181, 222–24, 323, 328, 363–64
as "Honest Abe," 45, 99, 354
"house divided" speech, *77*, 117
Illinois congressional races and, 34
in Illinois state legislature, 23
Lincoln-Douglas debates and, 13, 18, 19–20, 193, 223–25, 350

the "long nine" and, 152, *153*, *155*
married life, 27, 42, 97, 191
national renown, 18–20, 66, 147, 193
as New Salem store clerk, 45, 48
personal habits, 27, 28, 43, 82
playing town ball, 158
as presidential hopeful, 18, 44, 70, 96–97, 117–18, 362–65
as public figure, 27–28, 117–18
as public speaker, 28–29, 97, 255, 348–49, 350–53, 356, 366
Republican Party and, 96–97, 193
reputation of, 13, 22, 55, 76–77, 323–29
rescue of E. D. Baker, 50
"Scrap-book" of, 19
Senate race against Douglas, 22, 147
slavery issue and, 117, 181, 194, 222, 225, *278*
in Springfield, Ill., 22–28, 41–42, 49, 53, 58, 222, *326*
storytelling and humor, 28–29, 56, 323–25
strength of, 226, 265–66
voice of, 344
wit of, 54–55, 73, 114–15
—law career, 24–25, 35, 43–70
advice on entering the law, 48
Anderson murder case, 64–65
approach to a case, 194–95
approach to practicing law, 94–95, 255–56
Armstrong murder case, 66–69, 112, 148, 195, 328
as the best in Illinois, 57
Broadwell training with, 334
case opposing Logan, 205–6
choice of clothes, 57–58, 343–44
circuit riding and, 24–25, 54–55
courtroom manner, 56–57, 68, 72–73, 88–89, 108, 112, 147, 160, 254–55
Crafton training with, 33, 103, 165, 201, 351
criminal cases, 61
cross-examination, skill at, 165
damage claim against Springfield, 58

decision to go into the law, 46–47
Effie Afton trial, 18–19, 59, 328–30
essay on the honesty of lawyers, 45
esteemed by colleagues, 323–24
ethical standards for the law and, 60
on extemporaneous speaking, 334
fees and, 100–101, 327
Fleming v. Rogers & Crothers, 327–28
Fraim murder case, 63
Goings murder case, 29–30
Herndon as partner, 26, 53, 95, 191, *192*, 194
juries and, 56–57, 84–85, 323, 344–45
law offices of, 22, 26–27, 42, 49, 53, 95, 194–95, 200–201, 362, 368
legal education, 47–49, 55, 69, 228
letter to John King and, 255
license to practice law, 48–49
Logan and, 38–39, *38*, 50, 95–96, 99
Matson trial and, 44–45
meeting with clients, 101
as mentor, 334, 336, 344
number of cases tried by, 35, 70, *233*
partnership with Stuart, 49
precedents established by, 59–60
preparation by, 328–29, 343
principles of practice, 44, 225
railroad cases, 60, 95
record of wins, 60
reliance on memory vs. notes, 106, 123
rhetorical style of, 52
Shaw v. Snow Brothers, 349
summaries or closing arguments, skill in, 52, 56–7, 323–31, 344–47
Trailor brothers murder case, 61–62
Truett murder case, 50–53, 81
Wright case, closing argument, 325
Wyant murder case and, 65–66
Young Men's Lyceum speech on importance of law, 94–95
—*The State of Illinois v. "Peachy" Quinn Harrison*

activities following the verdict, 362
admission of hearsay evidence (Crafton's threats) and, 251–57, 292
admission of hearsay evidence (deathbed declarations) and, 81, 276, 284–87
analysis of the case, 111
Cartwright as old adversary, 33–34, 97–99, 103, 109, 268–69, 272
closing arguments and, 342–53, 356
complications of taking the case, 96–97
Crafton testimony, strategy for, 156, 160–61
cross-examination of prosecution witnesses, 134–35, 165–71, 177–78, 183
on defense team, 13, 18, 22, 34, 38–39, 74, 81, 96–101, 186
Grand Jury and, 41, 106–7, 107–10, 166
Henry testimony and, 177–78
initial interview with Harrison, 102–6
Judge Rice and, 72–73, 215
jury selection, 83–93
lack of objections, 124
Livergood testimony and, 125–26, 128, 357
Million testimony and, 120, 122–23, 124
objection to Epler testimony, 317, 317
Palmer's attack on, 353–55
questioning of defense witnesses, 237–40, 257–62, 292–93, 297–303, 308–15
questioning of Rev. Cartwright, 40, 81–82, 119–20, 269, 273, 277–82, 289–90
relationships with the principals, 27–28, 39, 44, 97–100, 103, 109, 165, 201, 351
relationships with the prosecution team, 74–75, 91, 110–12, 154, 190
self-defense strategy and, 12, 14, 32, 40–41, 44, 128, 195–200, 206, 341–42, 345–46
Shakespeare quoted by, 349, 351–52
timing of trial, 79

win and increased visibility, 365
Lincoln, Mary Todd, 27, 36, 42, 50, 97, 191, *192*, 201
Lincoln, Robert, 42
Lincoln, Thomas, 47
Lincoln-Douglas debates, 13, 18, 223–25
 Hitt's transcriptions of, 19–20, 147, 223–25, 350
 recognition of Lincoln and, 20, 224
Livergood, Peter, 187–89
Livergood, Silas, 107, 159, 161, 162, 304
 cross-examination of, 132–38
 testimony at trial, 125–39, 167, 357
Logan, Stephen Trigg, 38–41, *38*, 95–96, 228
 admissibility of hearsay evidence (Crafton's threats) and, 257, 292
 admissibility of hearsay evidence (deathbed declarations) and, 288
 Anderson murder case and, 65
 Bone testimony queried, 183
 Cartwright and, 277, 282
 closing arguments and, 337–42, 347
 as co-counsel, 38–39, 40–41, 83, 88, 147, 154, 186
 Crafton testimony and, 156
 cross-examination of prosecution witnesses, 132–38, 173, 185–86, 188–89
 Grand Jury hearing and, 106–7, 111
 interview with the defendant, 102–6
 on Lincoln, 367–68
 Lincoln and, 38–39, *38*, 50, 51, 95–96, 99, 205–6
 Lincoln's closing argument and, 353
 Lincoln's presidential candidacy and, 367
 Livergood testimony interrupted by, 126–27, 129
 Nottingham testimony query and, 311
 objection to Epler testimony, 317
 post-trial life and accomplishments, 367
 Purvines testimony and, 308, 309
 questioning of defense witnesses, 234–52, 259–66, 273–75, 277–82, 289–90

Trailor brothers murder case and, 62
Truett trial and, 50, 51
verdict and, 360
Wright case and, 325
Lukins, Peter, 47

Magna Carta, 144
Marshall, Humphrey, 324
Matheny, James, 61–62
Matson, Robert, 44–45
McClernand, John A., *78*, 107, 110, 204, 206–7, 212–13, 365, 368
McClure's Magazine, 289
Medill, Joseph, 19
Metzker, James, 67–69
Million, J. L., 31, 107, 120, 163, 185, 279
Crafton's deathbed declarations and, 40, 81, 109, 216
life after the trial, 369
testimony at trial, 120–24, 219, 319–20
Million, Mary Crafton, 369
Montez, Lola, 332
My Bondage and My Freedom (Douglass), 179

New Salem, Ill., 45, 47, 250
New York Tribune, 19, 366
Nicolay, John G., 22
Norris, James, 67–69
Nottingham, Abijah, 309–12
Nottingham, Colonel Jonathon, 309
Nukolls, Charles D., 86–87

Oliver Twist (Dickens), 283

Palmer, Benjamin, 211
Palmer, John M., 74–75, *75*, 112, 228, 354–55
admissibility of hearsay evidence (Crafton's threats) and, 251–57
admissibility of hearsay evidence (deathbed declarations) and, 275–76, 288
antislavery views of, 367
attack on Lincoln during closing arguments, 353–55
belief in his duty as prosecutor, 211–12
children of, 211
as Civil War general, 367
closing arguments and, 353–56
Congressional run by, 364–65
cross-examination of defense witnesses, 240–48, 259, 262, 266–67, 289–93, 297–301, 303–15
death of son, Benjamin, 211

jury selection and, 83–84, 87, 92
Lincoln and, 74–75, *75*, 111–12, 353–55, 364–65
Lincoln's praise of, 347
Lincoln's presidential candidacy and, 365, 366
Lincoln's strategy and, 205–8
murder charge vs. manslaughter, 110, 209–10
objection to Lincoln's motion, 358
post-trial life and accomplishments, 367
prosecution team and, 74–75, 80, 110–11, 154–77, 204–16, 361
questioning of prosecution witnesses, 120–32, 154–56, 171–77, 181–88
rebuttal case, 315–20
strategy for Cartwright's testimony, 212–16
strategy for Harrison case, 111–12, 191, 205, 208–16
work habits, 204

Patterson, William, 88
Payton, Isaac, 87
Pickrell, M. H., 92
Pierce, Jefferson, 91
Pilcher, Moses, 87
Pirkins, Joseph B., 146, 151
Pitman, Sir Isaac, 21
Pleasant Plains, Ill., 30, 32, 109, 316
aftermath of the trial and, 364
Atherton farm and stores in, 263
Crafton family in, 120, 159, 184
founder Jacob Epler, 216, 316
Harnett as resident in, 172
Harrison family in, 120
Livergood family in, 125, 187
Million as doctor in, 120, 131, 216
Short & Hart drugstore, 30, 79, 121, 234, 251
Turley and, 300–301

Polk, John Knox, 181
Prentiss, Sargent, 324
Purviance, John, 186
Purvines, John, 307–9
Purvines, William, 186–87, 260, 307

Quincy, Samuel, 215

railroads
Alton Express from Chicago, 17, 232

Effie Afton trial and, 18–19, 59, 328–30
grisly accidents and, 122
Haring v. New York and Erie Rr. Co., 122
Illinois Central Railroad, 95
Interstate Commerce Act of 1887 and, 368
Lincoln law suits and, 60, 95
rapidity of, 364
sleeping chairs installed on, 232
Raleigh, Sir Walter, 253
Ray, Charles H., 19
Republican Party/ Republicans, 24, 33
convention, 1860, 70, *78*, 366–67
Harrison trial jurors and, 86
Lincoln and, 70, 96–97, 117–18, 194, 365
Lincoln supporters, 366
Palmer and, 74, 365
slavery and, 97, 118, 194
Rice, Edward Y., 72–73, 81, 82, 243, 301
admissibility of hearsay evidence (Crafton's threats) and, 251–57, 292
admissibility of hearsay evidence (deathbed declarations) and, 215, 275–76, 282–89, 317
aftermath of the trial and, 363
Cartwright questioned by Lincoln and, 275–76
closing arguments and, 333, 342
courtroom decorum and, 127, 149, 229, 268, 282, 317, 360
courtroom temperature and, 228, 332
Crafton testimony and, 159, 165
jury instructions, 320–21
jury selection and, 92
Lincoln addresses, 135–36
Lincoln's outburst and, 284–87
Lincoln's praise of, 347–48
Lincoln's relationship with, 72–73, 215
Livergood testimony and, 132
quotes Dickens, 283
quotes Shakespeare, 283
post-trial life and accomplishments, 368–69
reading of the verdict, 360
robe for, 228–29

rulings on Epler and Million testimony, 320
testimony and, 159, 189
trial begins, 79
trial ending and adjournment, 361
river transportation, 58
Effie Afton trial and, 18–19, 59, 328–30
Robbins, Silas W., 73
Robinson, George, 88
Rusk, Anson, 65

St. Louis Republican, 65
Sangamon County, Ill., 24
Courthouse, 44, 71, 73, 78–79, 86, *233*
Grand Jury, 32
Scates, Walter, 58
self-defense, law of
in Boston Massacre trial, 207–9
in cases of manslaughter, 200, 209
Commonwealth v. Selfridge and, 199–200
in frontier America, 197–98
Grainger v. State, 202
Harrison murder case and, 128, 136, 194–200, 206, 209, 341
history of, 195–96
in Illinois, 198
Selfridge, T. O., 199–200
Seward, William H., 118, 366
Shakespeare, William, 283, 349, 351–52
Shaw, J. Henry, 328
Shaw v. Snow Brothers, 349
Shields, James, 36–37
Short, Benjamin, 30–31, 165, 318
charged as accessory, 364
testimony at trial, 234–49, 347
witnesses testify about the fight and, 126, 128, 131, 133–34, 160–61, 164, 167–68, 171
Slater, Rev. John, 109, 297–99
slavery, 179–81, 194
antislavery movement, 117
Cartwright's antislavery views, 98, *278*
Chase's position on, 97
colonization solution and, 181
as divisive issue, 117
Dred Scott decision, 117
Hitt's position, 181, 221
Lincoln-Douglas debates and, 19, 225
Lincoln's "house

divided" speech and, 117
Lincoln's positions, 117, 181, 194, 222, 225, 269
Republican position on, 118, 194
southern Illinois (Egypt) and, 223
Thomas on the South's position, 221–22
underground railroad and, 24
Speed, Joshua, 28, 49, 53, 349
Springfield, Ill., 14, 17, 22–26, *77*, *78*, 217
Anderson murder case, 64
capitol building in, *77*
Diller's Drug Store, 191
Donner-Reed Party and, 146–47
Globe Tavern, 41–42, 43, 71, 112–15, 191, 216–19, 364
Harrison murder trial and, 71, 118–19, 332
icehouse in, 210–11
Lincoln and, 22–28, 41–42, 49, 53, 58, 222, *326*
Lincoln's law offices, 22, 26–27, 42, 49, 53, 95, 194, 200–201, 362, 368
the "long nine" and, 152, *153*, *155*
Montez lectures in, 332
presidential library in, 15
railway station, 17, *23*
Richelieu performed in, 332
Sangamon County Courthouse in, 44, 71, 73, 78–79, 86, *233*
Speed's general store, 28, 49, 53
state capital moved to, 153, *153*
Trailor brothers murder case, 61–62
train transportation to, 17, *23*, 356, 364
underground railroad in, 24
water closets and outhouses, 296
Young Men's Lyceum, 94
Stanton, Edwin M., 57
State Democrat, 217
State of Illinois v. "Peachy" Quinn Harrison, The, 12
admissibility of hearsay evidence (deathbed declarations), 34–35, 214, 275–76, 282–89, 317
admissibility of hearsay evidence (threats by

Crafton), 210, 251–57, 291–92
copies of the transcript sold, 20
coroner's inquest, 41, 81
courtroom temperature and, 78, 178, 190, 240, 256, 267, 292, 337, 339, 357
defense closing argument, 335–53
defense strategy, self-defense, 32, 40–41, 44, 80–81, 128–29, 133–38, 156–57, 168–69, 171, 177, 183–84, 197–98, 209, 238–39, 245–48, 252, 261, 288, 291, 300–315, 341, 345–46
defense team (Cullom, Herndon, Lincoln, Logan), 13, 18, 22, 37–39, *38*, 44, 74, 80–82, 87–88, 91, 107, 133–34, 148, 156, 188, 190, 228, 233, 361
defense witnesses, 109, 233–49, 257–82, 289–316
ending of the case and adjournment, 361
facts of the case, 30–32, 35–37, 104–6, 109
as fair trial, 73, 211–12
Grand Jury indictment, 41, 106, 107–10
Hitt hired to create a transcript of, 20
judge, Edward Y. Rice, 72–73, 81, 82, 135–36
jury deliberation, 358–60
jury selection, 83–93
lawyers' clothes and, 333
Lincoln's impassioned outburst, 284–87
Lincoln's motion on Livergood testimony, 357–58
Lincoln's win and increased visibility, 365
Motion for Sequestration, 119
murder charge vs. manslaughter, 110, 209
opening of trial, 71–93
prosecution's closing argument, 333–35
prosecution's position, 41, 80, 107
prosecution's strategy, 111–12, 191, 205, 216
prosecution team (Broadwell, Palmer, and White), 74–75, *75*, *78*, 80, 91, 107, 110–11, 154–77, 204–16, 361

prosecution witnesses, 107–9, 120–39, 154–78, 181–88, 319–20, 357
public interest in, 15, 32, 69–70, 71, 110, 118–19, 217
rebuttal case, 315–20
spectators at, 76, 90, 119, 139, 149, 232, 238, 243, 287, 315, 348
timing of trial, 79
transcript discovered, 13, 372
verdict, 360
witnesses, number of, 40, 109
See also Hitt, R. Roberts; Lincoln, Abraham; Palmer, John; Rice, Edward Y.; *specific lawyers and witnesses*
Stevens, Alexander, 210–11
Story, Joseph, 48
Stuart, John Todd, 49, 65
Swett, Leonard, 65, 84, 324

telegraph, 230–31, 332
transmission of trial transcript and, 230, 321, 332, 364
testis (Latin), 307
Thomas, James, 115, 218–27
Thompson, William B., 361
Trailor brothers, 61–62
Treat, Judge, 146
Treatise On Pleading And Parties to Action (Chitty), 48
Treatise on the Law of Evidence (Greenleaf), 48
Tremont Hotel, Boston, 296
Truett, Henry, 50–53, 81
Trumbull, Lyman, 193
Tuft, Otis, and Otis elevator, 116
Turley, Tom, 300–301, 304
Twain, Mark, 83

United States
description of, 1859, 116–17
indoor plumbing and, 296
Lincoln's "house divided" speech and, 117
magazines of the nineteenth century, 141
national mail service, 220
as a nation governed by laws, 142
presidential election, 1860, 117–18
slavery issue in, 19, 24, 97, 98, 117, 179–80, 194, 221–25, *278*

solar storm (1859), 231–32
town ball (later baseball), 204
See also law in America; railroads; telegraph
U.S. Constitution, jury trial and, 142–43
U.S. Supreme Court, *Dred Scott* decision, 117

Vandalia, Ill., 23

Ware, James, 304
Wells Fargo, 220, 226
Whig Party, 34, 87, 194
White, Horace, 19
White, James B., 76, 107, 204, 206, 209, 212, 213
White, Skinny Thomas, 259–62
Whitney, Henry, 26
Wilson, Robert L., 23
Winship, George, 266
Wolfe, Solomon, 217
Works of John Adams Esq, Second President of the United States, The (Adams), 207
Wright, Cyrus, 115
Wyant, Isaac, 65–66
Yates, Richard, 193–94

Zane, James S., 312–15

QUESTIONS FOR DISCUSSION

1. We all come to any book or story with preconceived notions. How did this book either support or change your perceptions about Abraham Lincoln?

2. In addition to indulging in the curiosity factor of Lincoln's own words spoken in the courtroom in defense of a killer, the authors wanted to tell the story of the developing American legal system. Self-defense has been an essential aspect of American jurisprudence since its inception.

More than 150 years later, would a jury today be more or less sympathetic to the defense depicted in this courtroom?

3. The authors write, “Lincoln might not have been able to cite the statutes verbatim, but he understood the spirit of the law; he knew that virtue was supposed to be rewarded and wrongdoing should be punished.” Yet in at least one case cited in the book he helped a woman who supposedly killed her husband escape judgment by the legal system. Was Lincoln right in his belief? Was he right in his actions?

4. The key testimony in this case was a deathbed “confession” from the victim, taken and then testified to by the accused killer’s grandfather, Reverend Cartwright. Should he have been allowed to testify for his grandson? Should the prosecution have made a larger issue of his relationship to the defendant?

5. Lincoln spent the early part of his career “riding the circuit,” bringing the benefits of law to the frontier. This is how the law in America developed. How important do you believe this frontier justice proved to be in American history, and do you feel early Americans had more faith in the law than we do today?

6. Lincoln is described during the trial going “over

the top of the bench on top of the judge." Do you recall ever reading about Lincoln being so animated? Did this overt display of emotion surprise you? Do you believe this was an honest reaction or theatrics? If theatrics, do you think he was wrong to do this?

7. All of the participants in the trial knew and respected each other and often worked together. In fact, several of them received appointments during the Civil War. Thinking of the way the legal system works today, in which opposing counsel seem to be enemies, do you think this kind of camaraderie would be helpful to the system?

8. What broad traits exhibited by Lincoln in this trial became more pronounced during his presidency? Are there aspects of his character on display here that might have proved helpful if emphasized during his presidency?

9. The American legal system was based on British law, but out of necessity it took on a very American character. From what you read in this book, what aspects of this pioneer application of the law are still visible in our legal system today?

10. Was justice done in this trial?

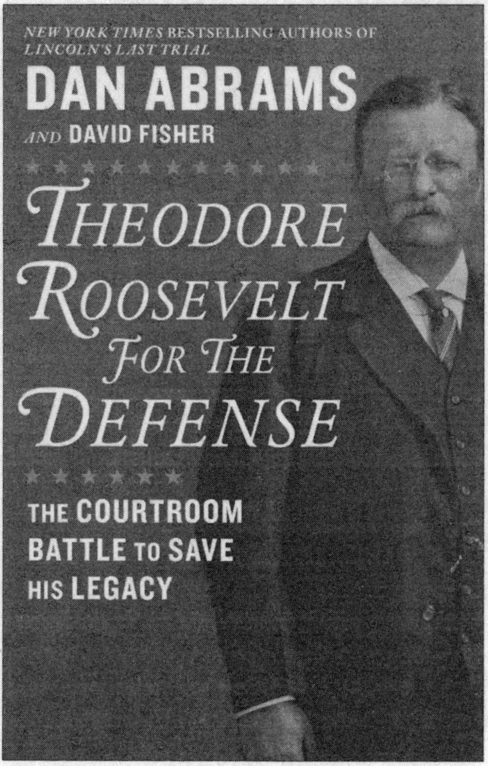

READ MORE THRILLING HISTORY FROM DAN ABRAMS AND DAVID FISHER.

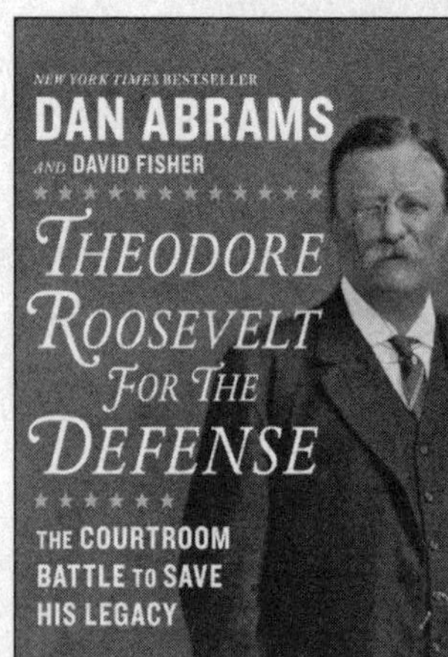

NEW YORK TIMES BESTSELLING AUTHORS
DAN ABRAMS
AND DAVID FISHER
JOHN ADAMS UNDER FIRE
THE FOUNDING FATHER'S FIGHT FOR JUSTICE IN THE BOSTON MASSACRE MURDER TRIAL

"Makes you feel as if you are watching a live camera riveted on a courtroom."
—Diane Sawyer

HanoverSqPress.com
BookClubbish.com

HSPDA191Tall